Nets and Coracles

Repairing a net at Appledore, Devon, circa 1900

Nets and Coracles

J. Geraint Jenkins

Keeper of Material Culture
Welsh Folk Museum

Illustrated by Peter Brears

David & Charles

Newton Abbot London

North Pomfret (VT) Vancouver

International Standard Book Number 0 7153 6546 0
Library of Congress Catalog Card Number 74-81005

Set in IBM 11/13pt Journal
and printed in Great Britain by
Redwood Burn Limited Trowbridge Wiltshire
for David & Charles (Holdings) Limited
South Devon House Newton Abbot Devon

Published in the United States of America
by David & Charles Inc North Pomfret
Vermont 05053 USA

Published in Canada by Douglas David & Charles Limited
3645 McKechnie Drive West Vancouver BC

Contents

List of Illustrations

PLATES

FIGURES IN TEXT

Hel wyf heolau'r afon
Ar hyd dwr a dryll a rhwyd don
Ei dwy ffon ydyw ei ffyniant
A'i diwyd waith a'i dau dant.
Ai gwydrau blwm i'w godrau
A thramwy hon i'w thrymhau

Dwr hardd a dorrir a hi
Dwr a dreiddia dioer drwyddi.

MAREDUDD AP RHYS, c AD 1440

The boats which they employ in fishing or in crossing the rivers are made of twigs, not oblong nor pointed but almost round, or rather triangular, covered both within and without with raw hides; when a salmon thrown into one of these boats strikes it hard with its tail, he often oversets it, and endangers both the vessel and its navigator. The fishermen according to the custom of the country, in going to and from the rivers, carry these boats on their shoulders.

GIRALDUS DE BARRI, AD 1188

Preface

This book is an ethnological and historical study of traditional methods of river fishing that are, or were until recently, practised by professional fishermen in the estuaries and rivers of England and Wales. It is primarily concerned with the methods and techniques of catching migratory fish, namely the salmon, sea trout and eel, for most of the commercial fishing in fresh water today, and certainly all the trapping and illegal poaching, is concerned with the capture of these three species of fish. The work is not concerned with the already well-documented world of the angler and leisure fisherman, but is limited to those men whose livelihood was dependent on the harvest of the rivers. Many use methods and gear that have persisted to this day with very little change for many centuries. Although the work describes methods of fishing for England and Wales generally, most of the detailed evidence presented originates in Wales and the rivers of the Border Counties. The main reason for this is that many traditional methods of salmon and eel capture, many of them dating back to prehistoric times, have persisted for much longer in the rivers of western Britain than anywhere else. Coracle fishing was hardly known outside Wales, while old-established techniques such as putcher, putt and stop-net fishing were limited to the Bristol Channel area. Nevertheless the book is not primarily a regional study, for such topics as eel capture and net making are described for the whole of England and Wales, while other peculiarly local techniques, such as the haaf-netting of north-west England and elvering in Gloucestershire, are described in detail.

I am grateful to the Fisheries Officers of the various River Authorities for their help and co-operation in this survey.

Amongst the many who helped me in my survey, I must give special mention to Lynn Davies of the University College of North Wales, Bangor, who, when he was on the staff of the Welsh Folk Museum, tape recorded the evidence of many fishermen and made the information he collected freely available to me. Raymond Rees, a coracle fisherman of Carmarthen, provided me with a great deal of priceless information and spent many hours writing the history and traditions of his fraternity for the first time ever. Above all, I owe a special debt of gratitude to the dozens of river fishermen that gave freely of their time to talk to me. Many of them allowed me to accompany them on fishing expeditions and all received me with courtesy and hospitality. It is to them, the netsmen and coracle men of England and Wales, that I dedicate this volume.

St Fagans, 1973 J. GERAINT JENKINS

1 Introduction

I

As befits a country with a long indented coastline, numerous swiftly flowing streams and broad river estuaries, the fishing industry has been well established in Wales for many centuries. Fishing for a wide range of freshwater and sea fish as well as shellfish has provided a livelihood for many generations of riverside and coastal dwellers. In the twelfth century, for example, Giraldus Cambrensis[1] described the abundance of salmon and trout in the Wye and Usk, and mentioned 'the noble River Teivi that abounds more than any river of Wales with the finest salmon'. 'The fishinge of Pembrokeshire', said George Owen of Henllys in 1603,[2] 'is one of the cheefest worldlie commodities, wherewithall god hath blessed this countrey, which fishinge are of diverse sortes, taken at diverse tymes of the yeare and that at diverse places.' Salmon, eels, crabs, oysters, herrings, as well as many other species, were taken in plenty, and fishing was undoubtedly one of the main pursuits of south-west Wales in the seventeenth century.

Travellers to Wales were all impressed with the abundance of fish in Welsh rivers and along the coast. Writing of Cardiff in the first decade of the nineteenth century, for example, Donovan[3] was impressed with the quantity of sea trout taken from the River Taff, where he 'observed fishermen taking them in vast numbers in their nets'. Herring fishing, too, was practised in all parts of Wales, but especially in Cardigan Bay and along the north Caernarvonshire coast. Herring fishing, salting and exporting made a substantial contribution to the economy of Wales until at least the end of the nineteenth century and 'the brigades of herrings' that Pennant described in 1776[4] were caught in vast numbers by full-time fishermen as well as by part-time fishermen, part-time farmers that were commonplace in coastal communities. Many villages such as Aber-porth in Cardiganshire, Llangwm in Pembrokeshire and Nevin in Caernarvonshire were almost exclusively fishing villages, and the economy of many coastal towns, such are Aberystwyth, Milford Haven and Pwllheli, leaned very heavily on the fishing industry.

Since the fishing industry has been well established in Wales for many centuries, it is only natural that fishermen should have developed their own way of life, their own customs and their own rights and privileges relating to fishing. Many of them lived in closely-knit communities that had very little contact with the surrounding countryside. Take, for example, the coastal villages bordering Cardigan Bay. In the past, before the advent of modern means of transportation had overcome many of the obstacles to movement presented by natural barriers, the seaboard villages were, in effect, self-contained localities which developed distinctive characteristics of their own. The Cardiganshire village of Llangrannog, for instance, is located at the mouth of a deep gorge-like valley, with sharply rising gorse and bracken-clad hills swelling up on both flanks of the village, obstructing the inhabitants' view of the land, and cutting down to a minimum the degree of contact between the village and the surrounding countryside. The inhabitants had no outlook but the Irish Sea, and by tradition the sea plays and has played an important part in the economic and social life of this village.[5]

The strong sense of community that was so characteristic of the fishing communities was by no means limited to coastal villages, for it was also found among riverside dwellers as well. The coracle fishermen of West Wales, for example, were described by the Commissioners on Salmon Fisheries in 1861[6] as 'a numerous class, bound together by a strong *esprit de corps,* and from long and undisputed enjoyment of their peculiar mode of fishing, have come to look on their river almost as their own, and to regard with extreme jealousy any sign of interference with what they consider their rights'. Coracle fishermen, therefore, formed almost a closed community that was impossible for a stranger to enter, and the fraternity administered a river through set rules of privilege and rights that had been passed down from time immemorial. They were oral laws passed down from father to son in closely knit, riverside communities.

On many rivers fishing is a family tradition and methods of fishing have been passed down from generation to generation over many centuries. On the Usk, for example, most of the drift netsmen who operate from Newport are members of the Sully family, but the most outstanding example of family tradition is that of the Bithell family, who fish for salmon in the Dee estuary. The Bithells, who all live in one particular section of the town of Flint, have been salmon fishermen for at least three hundred years, and they use a type of net, the drifted trammel, that has virtually disappeared from all other parts of Britain.

Seine netting is a universal method of fishing in estuarine waters, but although shore seines are similar in general design throughout the country, there are many local variations in the dimensions and technique of using the net. There are local rules relating to the allocation of fishing stations: there are local omens denoting a good harvest and equally potent omens that promise nothing but failure. It seems that custom and belief and tradition play an important part in the life of all fishing communities, and although these communities are considerably smaller than they were, methods of selecting fishing grounds and the rights and privileges associated with fishing have changed but little since professional fishing was amongst the most important of rural pursuits.

II

Most of the commercial fishing and certainly all the trapping and illegal poaching in Welsh rivers today is concerned with catching salmon *(salmo salar)*[7] and its close relative, the migratory trout or sewin *(salmo trutta)*.[8]

Adult salmon ascend the rivers, mainly in the summer months, to the gravelly shallows or 'redds' *(maran,* pl *maranedd, cladd* or *gwely)*, where breeding takes place between September and February. The salmon does not feed at all in fresh water, and, as a result, by the time they reach the spawning grounds, they are in a poor condition. A certain amount of illegal poaching is carried out at this stage in the salmon's life, mainly to obtain the roe of the hen salmon, once widely used as bait for trout. A hen salmon *(hwyfell, hwyddell* or *chwiwell)* lays between eight and nine hundred eggs for every pound of her weight and the eggs are fertilised by the cock salmon *(cemyw)* before being covered with gravel. In the spring the eggs are hatched and the tiny salmon, or 'alevins' *(silod y gro, silod y gôg, brith y gro* or *silod brithion),* are provided with a yolk sack, which is gradually absorbed by the growing fish. The adult salmon, after spawning, return to the sea as 'kelts'; many will die before they reach the sea; many will return to the spawning grounds in later years.

Meanwhile the alevins grow and begin to feed in the river. The alevin becomes a 'parr' *(eginyn)*, a trout-like fish, with a series of dusky mauve marks along its sides. As a parr it remains in the river for a year or more and gradually changes its colour to silver. The parrs then congregate in considerable numbers, when they are known as 'smolts', and they move towards the sea, where the major part of the growing and feeding is done. They sometimes return to the spawning grounds after a year at sea, and the 'grilse', as the one-year-old salmon is termed, weighs anything from four to nine pounds. Many will not return for two or more years and the fish are consequently much larger and heavier than the grilse. On rivers famous for large fish, such as the Wye, a salmon returning after four years at sea may weigh as much as fifty pounds. In the years ending March 1969, 1970 and 1971, the weights of salmon caught by anglers, netsmen and trappers on the Wye are shown in Table 1 *(see page 17)*.

Within the present century, there has been a sharp decline both in the number of commercial fishermen and in the number of salmon caught. Nineteenth-century statistics, even if they are not completely accurate, indicate that phenomenal quantities of salmon were caught in Welsh rivers. On the Severn, for example, netsmen and operators of fishing traps caught 22,500 fish in 1870, 30,000 in 1883 and 20,950 in 1902. Even on a minor river like the Dwyfor in north Wales, the average catch in the eighteen-eighties amounted to over 1,300 fish annually, while on the Dee the annual catch varied from 12,000 to 16,000.

The annual catch on the various rivers of Wales during the present century (the figures relating to five years taken at random) is shown in Table 2 *(below)*.

TABLE 1

	Year Ending 31 March 1969	Year Ending 31 March 1970	Year Ending 31 March 1971
Under 7lb	467	548	573
7-10lb	968	901	1,414
10-15lb	1,454	1,158	1,960
15-20lb	1,133	959	881
20-25lb	433	289	212
25-30lb	91	36	41
Over 30lb	21	13	2
	4,567	3,904	5,083
Weight in lbs	59,308½	48,012	58,324

TABLE 2

River	Number of fish caught				
	1919	1929	1939	1949	1959
Conway	256	246	664	527	855
Dee	No information	4,336	8,476	5,440	3,473
Dyfi[9]	No information	No information	393	910[10]	799
Severn[11]	15,500	10,300	3,410	5,590	5,127
Teifi	2,628	2,164	1,303	No information	1,313
Tywi & Taf	No information	371	1,095	1,672	547
Usk	1,555	1,979	2,498	1,918	1,571
Wye	4,455	5,751	3,978	4,420	3,961

During the 1970 fishing season, the quantity of salmon and sea trout caught in Welsh rivers is shown in Table 3 *(see page 18)*.

TABLE 3: TOTAL CATCHES OF SALMON AND SEA TROUT, 1970

RIVERS	SALMON				SEA TROUT			
	Nets & Fixed Engines		Rod & Line		Nets & Fixed Engines		Rod & Line	
	No	lb	No	lb	No	lb	No	lb
A. *Gwynedd River Authority* 1. Dyfi, Dysynni, Mawddach, Wnion, Eden, Artro, Dwyryd, Glaslyn	322	2346	1674	13503	675	2353	10250	10350
2. Dwyfor, Dwyfach & Rivers of South Caernarvonshire	116	977	81	662	302	272	4510	5042
3. Seiont, Gwyrfai & Llyfni	512	3034	447	2928	56	244	3281	4496
4. Ogwen, Aber & Rivers in Anglesey	148	815	190	1176	98	381	400	774
5. Conway & Lledr	136	936	807	5558	37	159	449	1122
B. *South West Wales River Authority*								
Teifi	825	7888	232	–	400	1303	Not available	
Tywi (& Taf)	377	3261	832	–	1214	3682		
Dau Cleddau	99	903	86	–	6	18		
Coastal	10	74			28	60		
Rheidol & Ystwyth			95					
C. *Usk River Authority*	1004	9877	729	6951	43	257	14	48
D. *Dee & Clwyd River Authority*								
Dee	1877	12946	884	8988¼	NONE			
Clwyd	1262	6258¾	140	1078¼	NONE			
E. *Glamorgan River Authority*	–	–	–	–			24	63

TABLE 3 *(continued)*

RIVERS	SALMON Nets & Fixed Engines		SALMON Rod & Line		SEA TROUT Nets & Fixed Engines		SEA TROUT Rod & Line	
	No	lb	No	lb	No	lb	No	lb
F. *Wye River Authority*	1127	10110	3955	48565	–	– NONE–		–
G. *Severn River Authority*	3440	35912	467	5908	–	– NONE–		–

The fishing for salmon is all important in Welsh rivers, for although other types of fish are present in the river, few attempts are made to catch them commercially. A few eel traps are found in such places as Llangorse lake in Brecknockshire, shad are taken with cleaching nets at Symonds Yat on the Wye and sparlings are caught by two seine nets on the Conway at Talycafn, but nowadays no effort is made to catch such fish as lampreys, lamperns or flounders, at least commercially.

III

In Wales, as elsewhere, commercial salmon fishing is carried on in waters where the right of fishing is public. Generally speaking, in coastal waters and estuaries everyone is entitled to fish and no one has an exclusive right. There are exceptions to this rule and private fisheries exist in several estuaries, the most notable in Wales being the Wye estuary, where the salmon fishing is the exclusive preserve of the Wye River Authority. In all the salmon rivers, the fishermen have to be licensed, and river authorities have the authority to limit the number of licences they issue.

In 1971 the following licences for commercial salmon fishing in Welsh rivers were issued:

1. SOUTH WEST WALES RIVER AUTHORITY
   - Tywi – 9 seine nets; 12 coracle nets
   - Taf – 1 wade net; 2 coracle nets
   - Daucleddau – 9 compass nets; 4 seine nets (barred from operation)
   - Coastal area – 1 seine net; 2 wade nets

Teifi – 6 coracle nets; 6 seine nets

2. WYE RIVER AUTHORITY

All netting and trapping were purchased by the Authority and all salmon fishing is carried out by employees of the Authority. These consist of 2 full-time employees, 7 temporary and part-time fishermen and 6 part-time lave netsmen. The methods of fishing are:

600 putcher traps at Beachley
5 stopping boats at Chepstow
1 drift net (tuck net) in estuary
6 lave nets
Putcher rank at Horse Pill (let)
1 eel trap at Llangorse lake

3. USK RIVER AUTHORITY

2 putcher weirs (total of 2,900 baskets) at Porton and Goldcliffe
1 putt weir (not in operation)
8 drift nets

4. SEVERN RIVER AUTHORITY

3 eel traps
43 eel nets and putcheons
2 eel night lines
35 lave nets (full season), 28 lave nets (half season)
4 seine nets
8 putts
4 putcher ranks (4,047 baskets)
3 leaders (hedging to putt or putcher ranks)

5. GWYNEDD RIVER AUTHORITY

33 seine net licences (3 Mawddach, 3 Glaslyn, 2 Dwyfor, 1 Daron, 2 Nevin, 4 Seiont, 2 Anglesey rivers, 2 Ogwen & Aber, 6 Conway, 8 Dyfi & Dysynni)
1 basket trap (Lledr)
1 weir (coop at Caerhun on Conway)

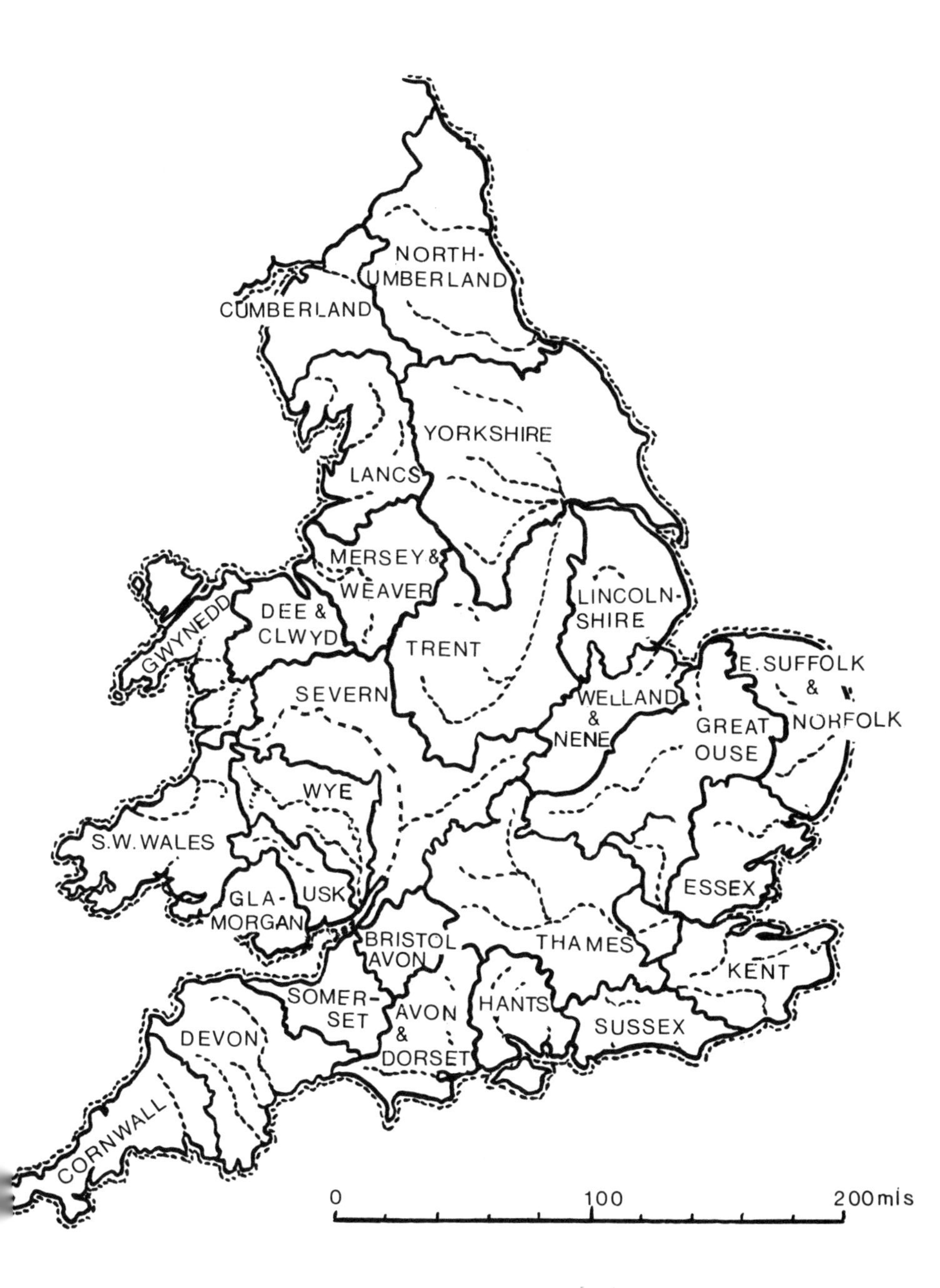

Areas of River Authorities 1972

6. DEE & CLWYD RIVER AUTHORITY
   4 trammel nets on Dee
   28 seine nets on Dee
   7 sling (drift) nets on Clwyd

7. GLAMORGAN RIVER AUTHORITY
   No commercial river fishing of any kind

On looking at England and Wales generally, it seems that there has been a decline in the number of people engaged commercially in salmon fishing. Rivers have been polluted so that, in general, salmon fishing is limited to the western coasts of Britain, and there is no salmon fishing south of Northumberland or east of Dorset. The number of licences issued by river authorities for salmon fishing in 1900 and 1955[12] is summed up in Table 4 *(see page 23).*

In 1970 the following methods of fishing were licensed by the various river authorities in England:

CORNWALL – 32 seine nets (3 Tavy, 14 Tamar, 2 Tavy/Lynher, 5 Tamar/Tavy, 3 Lynher, 4 Fowey, 1 Looe); 1 drift net (Camel).
CUMBERLAND – 1 seine net (Esk); 3 seine nets, 3 coops; approximately 200 haaf nets (tidal reaches of Solway); 1 coop (Derwent); 1 garth (Esk); 4 drift (hang) nets – 2 in western end of Solway Firth and 2 off West Cumberland coast.
DEVON – 2 salmon weirs; 80 seine nets.
DORSET – 6 (approximately) seine nets (Avon/Stow, Frome/Piddle).
LANCASHIRE – 18 drift (hang) nets (Ribble and Lune); 4 seine nets (Duddon); 46 haaf nets (Wyre); 14 lave nets (Kent and Leven).
NORTHUMBERLAND – 70 drift nets.
SOMERSET – 1 dip net; 1 putcher rank.

Appreciable quantities of salmon are caught in the rivers

of England and Wales and a summary of the total salmon catch during the 1971 season is given in Table 5 *(see page 24)*.

TABLE 4

RIVER AUTHORITY	1900		1955	
	Nets	Fixed Engines	Nets	Fixed Engines
Avon & Dorset	23	2	12	–
Bristol Avon & Somerset	3	245 (putchers)	42	550 (putchers)
Cornwall	25	–	38	–
Cumberland	142	3	117	5
Dee & Clwyd	66	–	49	–
Devon	85	5	83	2
E Suffolk & Norfolk	–	–	–	–
E Sussex	–	–	–	–
Essex	–	–	–	–
Glamorgan	2	–	1	–
Great Ouse	–	–	–	–
Gwynedd	31	4	32	2
Hampshire	–	–	1	1
Hull & E Yorkshire	–	–	–	–
Kent	–	–	–	–
Lancashire	59	1	74	1
Lincolnshire	–	–	–	–
Nene	–	–	–	–
Northumberland	143		54	
Severn	131	6457 (6130 putchers)	66	5574
South West Wales	124	–	65	8
Trent	1	–	2	–
Usk	1	3604 (putchers)	16	4 (licensed ranks)
Wear & Tees	71	–	9	–
Welland	–			
W Sussex	–			
Wye	20	1426 (1384 putchers)	11	9 (licensed ranks)
Yorkshire Ouse	54	–	36	–

TABLE 5
SUMMARY OF TOTAL SALMON CATCHES IN ENGLAND & WALES – 1971 SEASON

RIVER AUTHORITY	RODS											
	Jan	Feb	Mar	Apr	May	Jun	Jul	Aug	Sep	Oct	'X'	Total
Cumberland	82	97	103	24	27	42	113	513	252	399		1652
Lancashire											623	623
Dee & Clwyd			52	47	83	133	117	201	233	107		973
Gwynedd				12	28	275	221	1040	463	583		2622
SW Wales		9	40	41	98	414	153	715	227	84	37	1818
Usk		30	68	54	57	379	93	256	100	17		1054
Wye	2	249	472	490	701	1687	377	624	342	150		5094
Severn		62	82	84	109	201	73	125	45		12	793
Somerset												–
Devon		71	273	85	39	198	69	304	54	6		1099
Cornwall			24	18	48	235	176	317	163	53	272+	1306
Avon & Dorset		40	76	88	148	199	136	84	50		13	834
Hampshire		4	8	25	80	217	212	262	133	6	63	1010
Yorkshire						17	10	135	76	83		321
Northumbria		27	27	28	13	11	15	40	70	209		440
TOTALS	84	589	1225	996	1431	4008	1765	4616	2208	1697	1020	19639

'X' = Unallocated

\+ = 105 in November
154 in December
13 unallocated

TABLE 5 *(continued)*

RIVER AUTHORITY	NETS AND FIXED ENGINES									
	Feb	Mar	Apr	May	Jun	Jul	Aug	Sep	'X'	Total
Cumberland	22	21	33	108	300	893	608	4		1989
Lancashire			6	115	396	1470	1008			2995
Dee & Clwyd		51	101	281	775	1418	968			3594
Gwynedd									1293	1293
SW Wales		4	31	179	335	460	297			1306
Usk		5	22	375	638	604	371			2015
Wye		1	4	109	212	551	210			1087
Severn	20	54	146	611	838	1328	457			3454
Somerset				1	6	3	1			11
Devon	54	186	573	1449	1586	1893	989			6730
Cornwall		12	53	178	716	1774	1224			3957
Avon & Dorset	4	42	36	127	313	603				1125
Hampshire	1	1	3	9	30	138			1	183
Yorkshire			1	206	1265	3124	1369			5965
Northumbria									54201	54201
TOTALS	101	377	1009	3748	7410	14259	7502	4	55495	89905

ALL METHODS TOTAL:

Cumberland	3641
Lancashire	3618
Dee & Clwyd	4567
Gwynedd	3915
SW Wales	3124
Usk	3069
Wye	6181
Severn	4247
Somerset	11
Devon	7829
Cornwall	5263
Avon & Dorset	1959
Hampshire	1193
Yorkshire	6286
Northumbria	54641

IV

The history and development of inland fisheries administration from the earliest times is very largely concerned with salmon. This is because the salmon is, and always has been, a commercially valuable fish, and because the life cycle of the salmon is such that for the major part of its existence it is beyond purely private protection. It is for these reasons that Parliament has for a long time taken an interest in the salmon. 'There is a clause in Magna Carta about them, and the first Act, dealing specifically with salmon appears to have been passed in the year 1285, in the reign of Edward I; it imposed a penalty on taking salmon at certain times of the year. About a hundred years later in the reign of Edward III, other Acts were passed to regulate the construction of weirs, for although these are of great assistance in catching salmon, they can also entirely prevent their ascent of a river to the breeding grounds.'[13]

In succeeding centuries there was a proliferation of Acts relating to salmon fishing, one Act repeating earlier Acts with little chance of enforcement.

Until the Salmon Act of 1865, fishery law enforcement was entrusted to overseers or conservators appointed by the Justices of the Peace. 'The duties of these officers were undefined; they were unpaid and their powers were extremely limited and the system of nomination was lax. As a result there was much difficulty in carrying the various Acts into execution. Another defect of the law was that each county, instead of having regard to the general welfare of the whole river, was concerned only with those parts which flowed within its own boundaries.'[14]

As time went on, the law relating to salmon fishing became confused and uncertain, with the result that in 1860 a Royal Commission was appointed to inquire into salmon fisheries. This report[15] provides priceless information on the condition of salmon fishing in England and Wales at that time. As a result, the important Salmon Fisheries Act was enacted in 1861, and this Act has been the basis of all fishing legislation since. 'This', says the 1961 Report,[16] 'gave a reasonable and at that time, up to date code of law' but 'It still, however, provided for authorities which for the most part had no jurisdiction over complete river systems and left enforcement of the law to persons appointed by the Justices of the Peace. The Act went a little further than its predecessors in placing the general superintendence of the salmon fisheries throughout England and Wales under the central control of the Home Office, whose duties in this respect were transferred to the Board of Trade in 1896, and to the Board of Agriculture and Fisheries in 1903.'

Although the 1861 Act was a step forward, it was only a small one because the area of jurisdiction of salmon conservators was still related to the needs of fishery regulation; nor were funds provided to enable them to carry out their duties. As might be expected, the system did not work satisfactorily, and in 1865 a new Salmon Fisheries Act was passed setting up Boards of Conservators with jurisdiction over so much of a river or group of rivers as the Secretary of State thought necessary for the proper management of the salmon fisheries. In most cases, of course, the area comprised the whole of one or more catchment areas, but where rivers or parts of rivers contained no salmon, they were excluded from the jurisdiction of the new boards and so, it appears, were rivers which were in a single ownership.

The Act of 1865 was important in other ways. In addition to being the first legislation to introduce the principle of the imposition and collection of licence duties, it also for the first time provided for the appointment to the local boards of persons directly representative of fishery interests. While Justices of the Peace were still authorised to appoint the conservators, provision was made for additional board members, termed *ex officio* members, who were either local Justices with property interests in the rivers or persons paying above a certain sum in licence duties. A later Act of 1873 modified the constitution of the boards by giving other owners and occupiers of fisheries, together with commercial net fishermen, a right to representation, and in 1888 the duty of nominating local government representatives was transferred from the Justices of the Peace to the county councils.

Under the 1865 Act, Boards of Conservators were provided solely for the management of salmon rivers. No boards could be set up for rivers which contained only trout or other freshwater fish. It was apparently assumed that the owners of such fisheries would protect their own property and had the power to do so. Between then and 1878 opinion seems to have changed. In 1878 there was passed the Freshwater Fisheries Act, which altered the law in two important respects. It empowered Boards of Conservators to be set up for waters containing trout or char but no salmon and, as a consequence, authorised the boards to charge licence duties for fishing for these fish. The Act also imposed a close season for freshwater fish, so bringing them for the first time within the scope of general legislation.

Continuing the same process, another Freshwater Fisheries Act passed in 1884 enabled Boards of Conservators to be set up for rivers which contained only freshwater fish and gave the boards the power to make bye-laws for the protection of these fish. The Act did not, however, empower the boards to levy licence duties for fishing for freshwater fish, so that those boards in areas where there were no salmon, trout or char had no means for raising revenue and, doubtless, in other areas there was strong objection to money provided by salmon and trout fishermen being spent for the benefit of coarse fish anglers.

It seems desirable at this point to give brief particulars of the number of fishery boards which had been formed. Under the Salmon Fishery Act of 1865 thirty-one fishery districts were defined and boards constituted for them. By 1894 this number had risen to fifty-three covering about three-quarters of England and Wales. The boards, especially those for the larger river systems, were apt to be heavily over-weighted with local authority representatives and to have on them insufficient members directly interested in fishery management and conservation. On the recommendation of a Royal Commission on Salmon Fisheries which was set up in 1900 under the chairmanship of Lord Elgin, the Board of Agriculture and Fisheries were, by the Salmon and Freshwater Fisheries Act, 1907, given power to constitute and regulate fishery boards and districts by means of

Provisional Orders. The Act also contained an important provision which enabled fishery boards to charge licence duties for freshwater fish. By the year 1923 many fishery boards, particularly those for the larger districts, had made use of these powers to remodel their constitutions, but there were still many who had failed to take advantage of the new procedure, perhaps because they found it cumbersome and expensive.

The Salmon and Freshwater Fisheries Act, 1923, cured this defect and consolidated all the earlier legislation. With minor amendments made in 1929 and 1935 it is still in force. Under the Act where a fishery district lay wholly within one county, the county council was allowed to appoint five members, and where the district extended over more than one county, each county council was entitled to appoint up to three members. Persons representing holders of rod and line licences for fish other than salmon were given seats on the boards in addition to the earlier representation of fishing interests. Some of the anomalies of the law were thus removed, and coarse fish were, to a large extent, given the same protection as were salmon and trout. Where the Act gave less than full protection it enabled the fishery boards to obtain power by Ministerial order to extend the provisions to cover freshwater fish and, in return, to levy a licence duty for fishing for those fish.

By 1948 the number of effective fishery districts had fallen to forty-five, by reason of amalgamation of areas and because a number of the smaller boards had found it impossible to continue their operations as the result of rising costs. Some of the boards still functioning had incomes of the order of £200 a year or less, which was hardly sufficient to meet the cost of an efficient organisation. The setting up of river boards under the River Boards Act, 1948, introduced an entirely new factor into local fisheries administration by placing each river system in England and Wales under the charge of an authority responsible for the unified control of salmon, trout and freshwater fisheries, land drainage and the prevention of river pollution. Under this Act there are now thirty-two river boards, which between them cover the whole of England and Wales apart from the Thames and Lee catchment areas and the environs of London.

The River Boards Act, 1948, provides that the expenses of the boards, so far as they are not defrayed out of other revenues, shall be met by precept upon the councils of counties and county boroughs in the river board area. The boards continue to charge for the issue of fishery licences, which often pay the whole or nearly the whole of fishery expenditure, and they have also power in certain circumstances to levy contributions on the owners or occupiers of private fisheries but their main source of revenue is derived from their precepts. The Act also provides that all the revenues of a river shall be available for defraying the expenses of the board generally. The membership of the boards in normal circumstances is limited by the Act to a total not exceeding forty, including one member appointed jointly by the Minister of Agriculture, Fisheries and Food to represent local fishery and land drainage interests. One consequence of these important and far-reaching changes was that fishery interests lost much of their hard-won representation on the earlier fishery authorities and, with land drainage interests, are now entitled only to a minor share of the membership of the boards.

Since 1948, legislation that has affected fisheries was the Water Resources Act of 1963 that was designed to give the Minister of Housing and Local Government the power to formulate a rational water policy designed to augment, redistribute and transfer water resources. The instruments of this policy were the newly constructed river authorities, which took over the various aspects of the use of water resources which up to this time had been the responsibility of a multiplicity of authorities. The Water Act of 1973 will delegate fishery duties to nine regional water development authorities in England and to a National Water Development Authority for Wales as from 1 April 1974. One of the duties of the new authorities will be the conservation of salmon and freshwater fisheries.

For notes to Chapter 1, see pages 305-6

2 Fishing Weirs and Stake Nets

Undoubtedly the trapping of fish is one of the oldest methods of fishing known to man and throughout the country, in rivers and along the coast, the remains of stone or wattle weirs may be seen, many of them erected centuries ago for the capture of salmon and other fish. In Wales the widespread occurrence of the word *gored* as well as other names associated with the artificial damning of the flow of water give some indication of the importance of weirs in the economic history of rural Wales.[1] 'The two most common names for weirs', says Melville Richards, 'are *argae* and *cored*.'[2] *Argae* does not always refer to fishing weirs for it could refer to a weir built in association with a mill, but the word *cored* on the other hand refers invariably to fishing weirs, especially those constructed of plaited withies. The Welsh word for a pool specifically named for fishing is *pysgodlyn* (*pysgod* = fish + *llyn* = pool) and there is evidence to suggest that pools of this type were commonplace in the Middle Ages, especially in association with the monasteries and great houses of Wales.

Some fishing weirs were constructed in such a way as to completely stop the mouth of narrow tidal creeks. On the ebb, water would run off, leaving the fish stranded behind the barrier of stone or brushwood. 'As soon as it was found that this barrier blocked the ingress of fish', says Davis,[3] 'some arrangement would be made to allow them to pass the barrier while the tide was flowing. This effect would be produced by an opening in the wall which would be closed at high water.' Examples of this type of weir survived until the nineteen-twenties in the 'fish ponds' of south Devon.

Another type of fishing weir that was extremely common

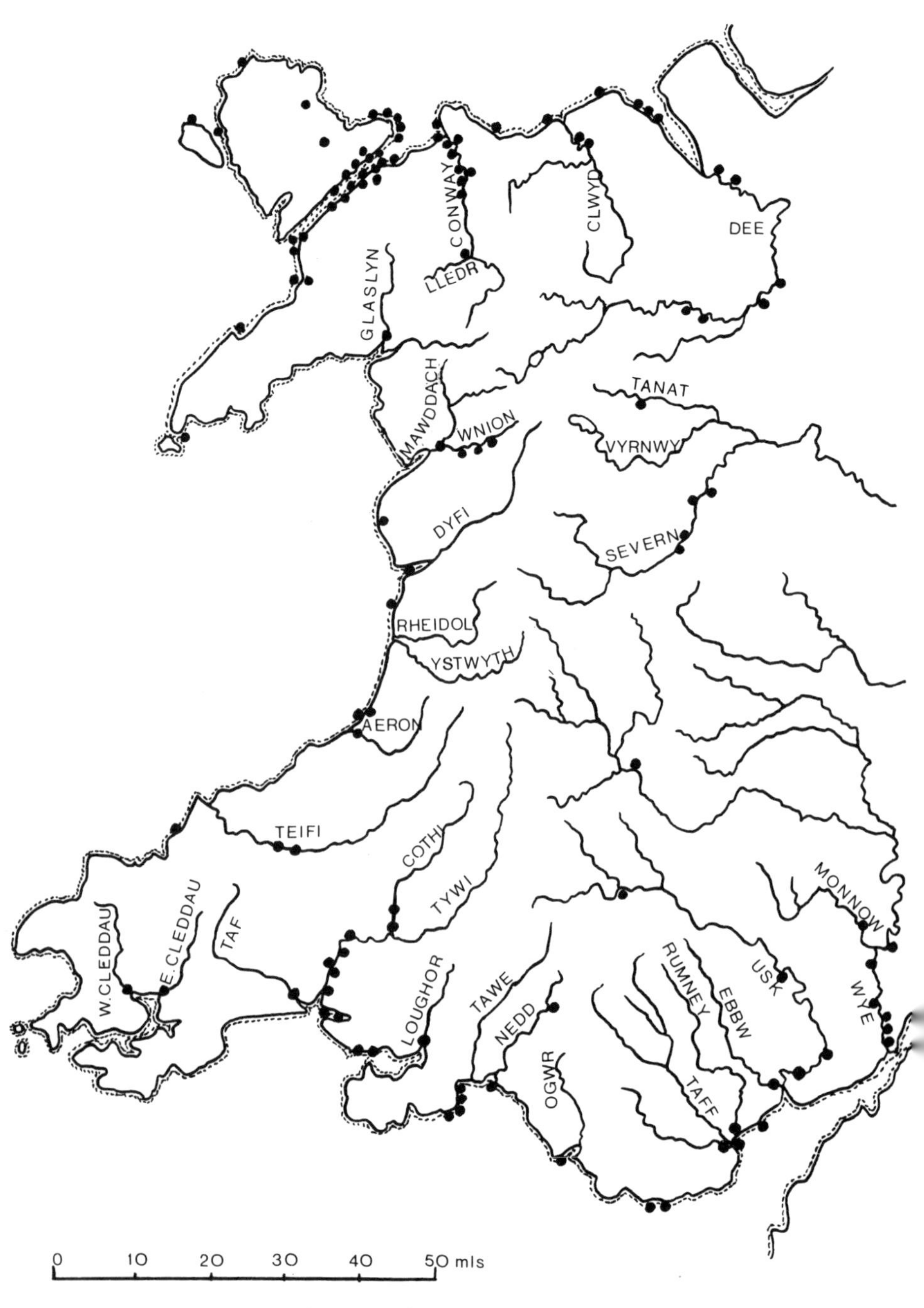

Distribution of known fishing weirs in Wales

before the Salmon and Freshwater Fisheries Acts of the eighteen-sixties was the type associated with the diversion of water to a mill or factory building. In some cases the weirs would be equipped with basket filter traps; in others salmon and other fish were speared or gaffed as they were stranded behind the walls of the weir. This type of weir was severely criticised by the Commissioners on Salmon Fisheries in the eighteen-sixties[4] as being 'of all the evils that affect the fisheries, artificial obstruction must beyond all question be regarded as the most pernicious . . . (They) offer great facility to the encouragement to the unfair means employed for destroying (the fish)'. According to the Commissioners, weirs associated with mills required 'to be vigorously dealt with. Of all the causes of injury to the rivers . . . none are more actively mischievous than the various modes practised for taking the salmon in connexion with mills and mill streams . . . We found that it was a very common practice to place baskets in the mill streams to catch the fry . . . and netting and other devices for killing salmon . . . But the worst advantage taken . . . has been the construction of fishing traps in connexion with their weirs or dams. We believe that these contrivances have been in most instances set up without any warrant or title; in blocking up the whole course of the rivers against the passage of the fish they form a standing violation of the principle if not the letter of the law'.[5]

The Salmon and Freshwater Fisheries Act of 1865[6] enacted that 'no fishing weir or fishing mill dam which was not lawfully in use on the sixth day of August eighteen hundred and sixty one, by virtue of a grant or charter or immemorial usage, shall be used for the purpose of taking or facilitating the taking of salmon or migratory trout'. A further restriction on weirs was demanded by the Act of 1923[7] which led to the abandonment of many fishing weirs. This Act states 'If a fishing weir extends more than halfway across a river at its lowest state of water, it may not be used for taking salmon or migratory trout unless it has a free gap of forty feet or not less than one tenth of the width of the river . . . situated in the deepest part, and level with the bed of the river and with sides parallel to the direction of flow. A fishing mill dam may not be used for taking salmon unless it has an approved fish pass.'

Not all weirs built across the flow of a river were associated with mills, for many, of ancient usage, were built especially for the capture of migratory fish. In Wales in the Middle Ages, for example, the Cistercian Order at Tintern Abbey had weirs on the Wye at Plumweir, Staweir and Alfred's Weir in the eleventh century. 'These were later exchanged (1148 +) for a share of Walwere, Halfwere and Badingsweir, but by 1224 Plumweir once again appears to have belonged to the house.'[8] They had other weirs on the Wye by 1330, at 'Britheksweir, Ishelsweir, Ashweir and Brocweir'. On the Severn they fished Alvredeston and Erlisgrove at Moat Grange. On the Usk they fished at Monkswood. Llantarnam Abbey also fished the Usk at Aberavon and Tredunnock, while Margam Abbey had weirs on the Afan and Neath. Neath Abbey had weirs on the Tawe, Neath and in Gower. 'Whitland had a weir in the Towy and was granted the fisheries of Haverfordwest every Friday and Friday night. Strata Marcella could fish at Aberystwyth and in the upper Dovey and Strata Florida . . . had weirs in the Ystwyth and fished at Aberarth and Penwedic. Cwmhir used Llyngwyn, while Cymer had rights in the upper estuary of the Mawddach . . . Aberconwy fished in the estuary of the Conwy right up to Aber-gyffin and also near the monastery and in the Dwyfach and Dwyfor rivers and the intervening coast where it constructed weirs . . . Valle Crucis had a weir at Overton.'[9]

According to George Owen[10] 'the greatest weare of all Wales' was on the River Teifi at Cilgerran, a weir 'built of strong tymber frames and artifically wrought therein with stones, crossing the whole ryver from side to side, having six slaughter places, wherein fish entring remaine enclosed, and are therein killed with an iron crooke, proper for that use'. This weir was also known to Giraldus Cambrensis[11] and it was 'considered to be one of the most profitable of the manorial rights and was invariably granted as a part and parcel of the Manor of Cilgerran . . . As early as the seventh year of the reign of Edward II a grant was made by the King to John de Hastings, who was Lord of

Cilgerran, of a certain weir . . . at the annual rent of twenty shillings'.[12] It seems that this weir flourished until the early eighteenth century, until another weir was constructed about a mile upstream at Castle Maelgwyn, Llechryd. This weir was granted by Queen Anne to the Coedmawr family and 'for a long period it proved a most productive speculation, and was let out to tenants from year to year at exceedingly high rents ranging between £80 and £150'.[13] The weir was destroyed by the Rebecca rioters in 1843. 'On the night of 18 July', says David Williams,[14] 'a crowd assembled to destroy it, but found that the soldiers had arrived before them. This weir was held under a lease of three lives by one of the local gentry, Abel Lewes Gower.' The weir was completely destroyed by the mob on 13 September 1843, despite the fact that Gower had employed a private police force to protect his weir. Other weirs in West Wales destroyed by the Rebecca rioters were those at Felingigfran on the Nevern and at Blackpool below Canaston Bridge in Pembrokeshire.

All these weirs, and they are but a few examples of the hundreds that must have existed on Welsh rivers before 1865, were those built across the flow of a river. There were other types of weir that were regarded as less injurious to the supply of salmon. At Gored Ddu on the Anglesey shore of the Menai Straits, for example, a weir was constructed between a small island and the mainland. Here the fish could pass round the outside of the island and towards the shore on the flood tide, but were held back by the wattled barrier of the weir as they made their way through the channel on the ebb. At Beaumaris, on the other hand, a weir known as 'Lyme Kilne Fishery'[15] was erected and in 1448 a second was built alongside it. The Beaumaris weirs were constructed of wattle with an arm at right angles to the main wall of the weir to provide a 'crew' or hook to prevent the egress of fish. A recess in the wall at the apex of the two arms could be closed or opened by a wooden trap door. The Beaumaris weir was operating in the nineteen-twenties and can still be seen near the lifeboat station, while another at Ferryman Warth, Beaumaris, has virtually disappeared.

In 1448 a second weir makes an appearance. The remains of two weirs can still be seen on the foreshore at Beaumaris.

The most common type of weir was that especially designed and constructed to fish along a coastline. 'The simplest barrier of this sort would be V-shaped, with its apex pointing in the direction of the ebb tide, so that all the fish coming inshore with the flood tide can easily pass the wedge-shaped obstruction, but with their return with the ebb, come between the arms and on the fall of the tide are gradually stranded in the apex . . . They are generally built of stone and stakes or hedging, each one being anything up to 400 yards in length and the apex being near low-water mark.'[16]

The Menai Straits and the shores of Cardigan Bay in particular had many fish weirs. In the Llanddewi-Aberarth district of central Cardiganshire, for example, it is said that there were about a dozen fish weirs operating in 1861;[17] by 1896 these had declined to nine and in 1924 only two weirs remained in operation.[18] All these were concentrated between the mouth of the Aeron and the mouth of the Arth, and they originally belonged to the monks of Strata Florida. 'The making of a *gored* (weir)', says Evelyn Lewis,[19] 'is a matter of strength rather than skill, for a strong wall of stones, taken from the beach and piled one upon another, is erected on the beach; the extension of the wall being usually about 200 yards or more. At the deepest point in the *gored* between two of the lower stones, there is an opening bridged by one very large stone supporting others. A drain is thus provided and across the drain are placed strong, slender stakes, or sometimes in their later days an iron grating.' Further up the coast there were *goredi* just north of Aberystwyth (Gored Wyddno);[20] at Tywyn, near the mouth of the Mawddach, and at Aberdaron, while in south Wales, Swansea Bay, especially the Oystermouth-Mumbles district, had as many as 13 stone-and-wattle weirs operating in the late nineteenth century.[21] These weirs were about six feet high and were composed of stakes driven into the sand and wattled so as to constitute a fence. In describing a Swansea Bay weir Matheson says,[22] 'The arms of the weir formed an angle of approximately ninety degrees and each measured up to 200 yards in length. At the junction of the arms, near low water mark, was a closely woven conical basket with its entrance facing the inner side of the weir. In order to retain some water when, during spring tides the tide ebbed beyond the limit of the weir, a layer of bushes and matting extended, close to the ground, along each arm of the weir, to a distance of about 50 yards from the basket.' Sometimes a continuous trap of two or three miles in length would be formed by a line of these weirs, the inner ends of the adjacent arms being in some cases united. Complaints about the destruction of young fish by these engines were frequent, and in some cases nets were substituted for the wattled fence, this being said to result in the capture of fewer small fish. Of course, the Swansea Bay weirs were not specifically designed for the capture of salmon and other migratory fish, but being fixed and solid structures they were able to trap all sizes and types of fish. The remains of the weirs may still be seen, for their construction was such that they were made durable by the silting of sand and stones around them. They were in use until the end of the nineteenth century, but were gradually supplanted by stake nets that persisted in Swansea Bay until the late nineteen-thirties.

In north Wales, especially in the Menai Straits, fishing weirs were commonplace and a weir at Penrhyn Castle, Bangor, near the mouth of the Ogwen, was in use in recent years. In this weir there was no special cage for the fish but they were caught in the apex made by the arms of the weir. In the Menai Straits, below the city of Bangor, the remains of Gored y Git may still be seen. It was referred to as early as 1588 and 'ended its career in 1852. It was leased in this year to Messrs. Daniel and Jonathan Russell, who designed it to utilize the mud flats as oyster beds. They removed the stakes and formed a series of banks of stone for the protection of their oysters . . . Their banks of stone . . . still remain.'[23] The *gored* was almost certainly the one referred to by the Rev John Evans when he toured north Wales in 1798[24] when he described it as 'a salmon snare'. It consists of posts driven into the ground at regular intervals with wattling in between. It was about 10 feet high and extended into the Straits in a semi-circle about 700 yards long.

Among other weirs in the Menai Straits were those at Borthwen Ferry House, at the mouth of the Cadnant and at Treborth. The last was located on the Caernarvonshire shore and was removed when the Britannia railway bridge was constructed. The most complete of all the weirs in the Straits is that of Ynys Gorad Goch, recently described in detail by David Senogles.[25] This is a double weir built from an island between the two bridges in the Menai Straits and it is capable of catching all types of fish. The fish are kept in the weir, not so much by the half circular shape of the arms but by the strength of the tide, and after a short time they are retained behind two 8-foot-high ramps at the inner end. Dr Ernest Benn of Menai Bridge, who has vast knowledge of the fish traps at Ynys Gorad Goch, describes their use as follows:[26]

'The two fish traps are enclosed spaces of about one-eighth of an acre each with a maximum depth of from eight to twelve feet on an average tide. They fill on the flood and run westward one hour before high tide by which time it is running fast, and it continues to run westward for about 6½ hours. It then changes direction and runs eastwards. It continues in this direction for almost 6 hours in all. The speed of the current is said to reach 7 to 8 knots in each direction on spring tides, and 5 to 6 on neaps.

'The funnel-shaped entrances to the fish traps open to the East and therefore channel the run of the ebb through the relatively narrow mouths of the traps which contain grills. These grills slope upwards to the West in the direction of the flow of the current, and further restrict the openings and accelerate the stream which carries with it any fish it contains. Once in the traps it is very difficult for fish to escape over the grills against the very considerable force of the stream. Most fish can only swim at speed for short bursts which are insufficient to overcome the speed and strength of the flow of water over the grills.

'As the tide falls, the water in the traps escapes. In the South trap the water percolates through the loose stone walls, but in the North trap, which had the loose stone walls replaced by solid walls in 1924, it runs out through iron gratings placed in apertures in the walls.

'In the lowest part of each trap there is a pool where the fish are left by the receding tide and where they are netted from stone walks built for the purpose.

'Whitebait can be netted by a single fisherman working either with a large ring net or with a simple landing net in each hand. He brings these together from each side of a shoal of whitebait as it swims by. To catch larger fish it is necessary for two men to operate a short seine net on weighted vertical poles. The fishermen start from a point at one end of a pool and work down opposite sides keeping the poles to which the net is attached close to the sides of the pool on which they are walking. When they reach the other end of the pool, fish are removed by hand nets.

'It is necessary to watch the falling tide to prevent birds making inroads on the fish before the fisherman is able to net them. Gulls by the hundred, terns and herons can empty a pool of whitebait in a comparatively short time, and herons can damage a catch of larger fish such as mackerel by wounding the fish even if they only succeed in lifting a few of them out of the water. It is necessary to open a sluice gate at the lowest place in each trap after each tide so as to allow any damaged or uncaught fish to return to the sea. If this is not done, some fish will be stranded or the birds will continue to attack the fish, and dead fish may contaminate the trap.

'The fish traps of Ynys Gorad Goch appear to be unique in that they depend on the speed of the current running over the grills to retain the fish in the traps until the tide falls below the level of the top of the grills. It is to this that they owed their success in the very large herring catches of the last century. The biggest catch which the writer himself has seen was 529 pollack on 30th August, 1953.'

The fish trap at Caerhun on the Conway above Talycafn is

one of the last to operate in England and Wales. This is specifically a salmon trap and belongs to a once important group of fishing devices known variously as 'boxes', 'cribs' or 'coops'. This method of fishing 'has the peculiar interest that it is extremely efficient and yet its use has greatly declined in recent years. Essentially this instrument consists of a weir with a gap in it. Gratings are set at an angle pointing upstream from each of the downstream corners of the gap, so set that the space between the gratings is only just sufficient to allow a salmon to pass through. The rush of water through the gap entices salmon to pass between these gratings, but their egress to the river above the weir is stopped by another grating set across the gap and they are trapped.'[27] At Caerhun the weir built about 5 yards from the bank of the Conway, runs parallel to the bank and is well over 100 yards in length. It is built of boulders and shingle and the water rushes through at a

Salmon coop at Caerhun on the River Conway, 1972

considerable speed. A wooden grating ensures that the fish are forced into the fish trap itself, the trap being a chamber measuring about 10 feet by 5 feet. Although local lore believes that the coop at Caerhun dates back to Roman times, it is unlikely that it was erected before about 1820. A witness to the Commission in 1861 stated that 'it was not there in my young days' and was set up 'about twenty or thirty years ago'.[28] But there is evidence to suggest that this weir, known as *Cored Hugh ap Thomas*, was known as long ago as 1539. At this spot the tide rises six and eight feet at spring tides and according to the 1861 enquiry many Llanrwst fishermen objected to the use of the salmon cage and the owner at the time, Hugh David Griffiths, said that when he set up the trap, around 1825, 'the Llanrwst people threatened to take it down' for, he adds, 'we used to catch on the average about seven or eight cwt . . . and as many as 15 cwt in a year'.

Although most fishing weirs, and indeed the walls of coops, are permanent, stoutly constructed structures, it was customary in some parts of the country, notably the Aberdaron district of Llŷn, to build small, temporary stone weirs that were dismantled at the end of the fishing season. Of course, with coops such as the Caerhun trap, it is essential that they can be put out of action during the closed season by the removal of the stakes that act as a barrier when fishing.[29]

Closely related to the fishing weirs are the seasonable fixed nets, usually staked, that are kept in position continuously during the fishing season but removed for the closed season. In many districts the stakes that support the net are kept in position throughout the whole year. Evidence of stake netting may be seen in such places as Carmarthen Bay and the Conway estuary, but until about 1940 the stronghold of stake netting was Swansea Bay, between Swansea itself and Mumbles Head. Here, says Davis,[30] 'are several large V-shaped nets locally known as "Stop Nets" or "Kettle Nets". They are the largest nets of the type found in the country, the total length of the two arms of each net amounting to about 700 yards. They are usually set in series with the tips of the arms of adjacent nets almost meeting, and with the axis of the net almost at right angles to the foreshore, the tips of the arms not reaching high water mark. The net is about 7 feet high, set on stakes and of 1 inch bar mesh. At the apex there is a circular, roofed cage about 12 yards in circumference, prolongations of the arms into which form the usual type of non-return trap. Local bye-laws require the cage to be in such a position that a pool is left in at low water.'

Until 1939, stake nets were far more common in river estuaries and along the Welsh coast, than they are today. Some of the nets, such as those used in Carmarthen Bay, were simple gill nets, 4 or 5 feet high and never more than 100 yards long, fixed to stakes and allowed to swim with the tide. The fish stopped by the net would be evenly distributed along the whole length of the set and would not be guided to one section as in the Swansea 'Kettle Net'. Davis describes the so-called 'Ferryside stop net' as being '100 yards long and 14 meshes deep, set in by the half. In some cases the end is twisted round . . . forming the "bung end" which functions like the inturning arms of fish weirs, of which it is an exaggerated form'.[31] Undoubtedly this has always been a characteristic of Carmarthen Bay stake nets, for an observer in 1863[32] states that most of the nets between Carmarthen and Llanelli were shaped like 'a ram's horn'.

A slightly more advanced method of setting stake nets was to set them in zig-zag fashion, to form a series of Vs, as at Llanfairfechan and Aber on the North Wales coast. Each arm of the V was about 50 yards long, but considerable variation in length could occur, according to the configuration of the coast at a particular point. In the Dee estuary, the stake nets were allowed to rise and fall freely with the tide. These nets could be simply single walls of gill netting or they could be fixed trammel nets. In the latter case, the nets that were widely used in the Connah's Quay district on the Dee consisted of three walls of netting – lint and armouring set across a width of river. The centre wall was of small mesh and loosely hung on the head-rope and attached to a lead line; the outer walls of coarser mesh were set tightly on the head and lead lines. In the Cardiff district until the nineteen-thirties, the so-called 'hang nets' were simple gill nets 300 feet long and 3 feet high, 'with the addition at intervals of butts -- cylinders supported on hoops, and having a non-return funnel to prevent the escape of fish'.[33]

By far the most complex of stake nets, that were widely used on the North Wales coast, from the mouth of the Conway to the Dee, was the so-called 'bag net' introduced into the area by Scottish fishermen in the eighteen-twenties. In the Conway estuary, for example, the Salmon Commissioners Report of 1861[34] notes that 'about twenty five years ago a party came' (to the Rhyl district) 'from the Solway Firth and put some stake nets here, and they took a great deal of salmon that way . . . The stake nets were cut one night by some ill-disposed persons'. In the Conway the so-called 'Scotch weirs' were introduced by Scotsmen in about 1820 and were regarded as a serious obstruction to navigation both on the Conway and the Dee.[35] The Scotch stell operated 'upon the principle of a leader running from or near high water mark seaward, again which the salmon, in their course along the coast, strike and, in their endeavour to find a passage, are guided into a narrow opening, the entrance to a chamber or trap, from which there is no escape. In some cases, these nets are of great extent, and have many chambers, the last being placed so far into the sea or channel as the very lowest tides will permit; it is never entirely dry and is generally waded into and fished with a scoup net, and to this chamber what is called a bag or fly net has of late been attached, which reaches still farther seaward . . . It appears they were first introduced on the Scottish side of the Solway Firth about the year 1780'.[36] The introduction of Scotch stells was objected to violently by the fishermen of North-east Wales for those in the Clwyd and Dee estuaries were destroyed by local people in the eighteen-fifties and they only remained on the Conway, where they were leased by the Corporation of the town to some immigrant fishermen. By 1870 they had been largely replaced by seine nets.

For notes to Chapter 2, see pages 306-7

3 Putchers, Putts and Basket Traps

Undoubtedly, the trapping of fish is one of the oldest methods of fishing known to man and throughout Wales, in rivers and along the sea shore, may be seen the remains of stone or wattle weirs (*goredi*), erected many centuries ago for the capture of salmon and other fish.[1] In addition to semi-permanent, stoutly constructed weirs, such as those in the Menai Straits, on the Conway and at the mouth of the Arth in Cardiganshire, there were also in all parts of Wales a large number of removable traps. These were known under a variety of names such as putts, putchers, baskets, cribs, hecks, crucks and inscale. Salmon fishing was forbidden in many of these by the Salmon Fishery Acts of 1861, 1865 and 1923,[2] except under grant or charter of 'immemorial usage'. By the Act of 1865, special commissioners were appointed to enquire into the legality of all 'fixed engines' for catching salmon. That such engines were established by grant, charter or immemorial usage had to be proved to the satisfaction of the commissioners who then issued a certificate of legality. The Act of 1865 states that 'any fixed engine which was in the use for taking salmon or migratory trout during the open season of eighteen hundred and sixty-one, in pursuance of ancient right or mode of fishing as lawfully exercised during that open season' could be legalised and allowed to continue. In this way the number, size and position of all weirs and traps became fixed for all time. Although many weirs have disappeared or become smaller in size within the last hundred years, no new weirs have been established, neither has the position of those in existence in 1861 been altered.

In Wales and the eastern borders, three types of removable basket traps are known. They are:

1. *The Putcher* This is the most common and widespread type that is found exclusively in the Severn estuary. Each putcher is a cone-shaped, openly woven willow or wire basket, five feet to six feet in length and about two feet in diameter at the mouth, tapering to about five-and-a-half inches in diameter at the tip. The putchers, which are used exclusively for salmon catching, are arranged in ranks of several hundred, ranged in three or four tiers to form a weir. They were widely used along the Monmouthshire and Gloucestershire banks of the Severn and are still used at Porton and Goldcliff in Monmouthshire. On the opposite side of the Bristol Channel in the estuary of the Parrett, putchers are known as 'butts'.

2. *Putts* These fixed engines, also known as 'putt nets', are less common today than they were in the past and are more complicated and carefully constructed than the putchers. Again, they have always been limited to the Severn estuary, and each putt is a closely woven basket trap, consisting of three sections. They are known as 'kipe', 'butt' and 'fore wheel'. The diameter of the kipe may range from as much as five feet at the mouth to fourteen inches at the tip. The butt which fits into the kipe gradually tapers to an approximate diameter of six inches, while the fore wheel *(voreel)*, which again fits onto the tip of the kipe, tapers to a diameter of about two inches. A removable wooden or straw bung covers the opening at the tip of the fore wheel. The putts are made of willow, hazel or whitethorn and according to the Clerk of the Usk Conservators in 1902[3] 'are like three buckets dovetailed one into another, and the last one is so close that nothing can get out of it; they can catch everything from shrimp to salmon'. Putts are used in single rows and a putt weir in the Severn estuary may contain as many as 120 baskets, with the mouths of the trumpet-like kipes facing the ebb tide, usually at the tail of a pool in the river.

3. *Basket traps* These are not used in weirs, but singly in rivers in north Wales. The only trap of this type in use in Wales at the present time is at Pwll-glas on the River Lledr, a tributary of the Conway near Betws-y-coed, Caernarvonshire. The trap is a conical-shaped hazel basket with a rectangular framework of

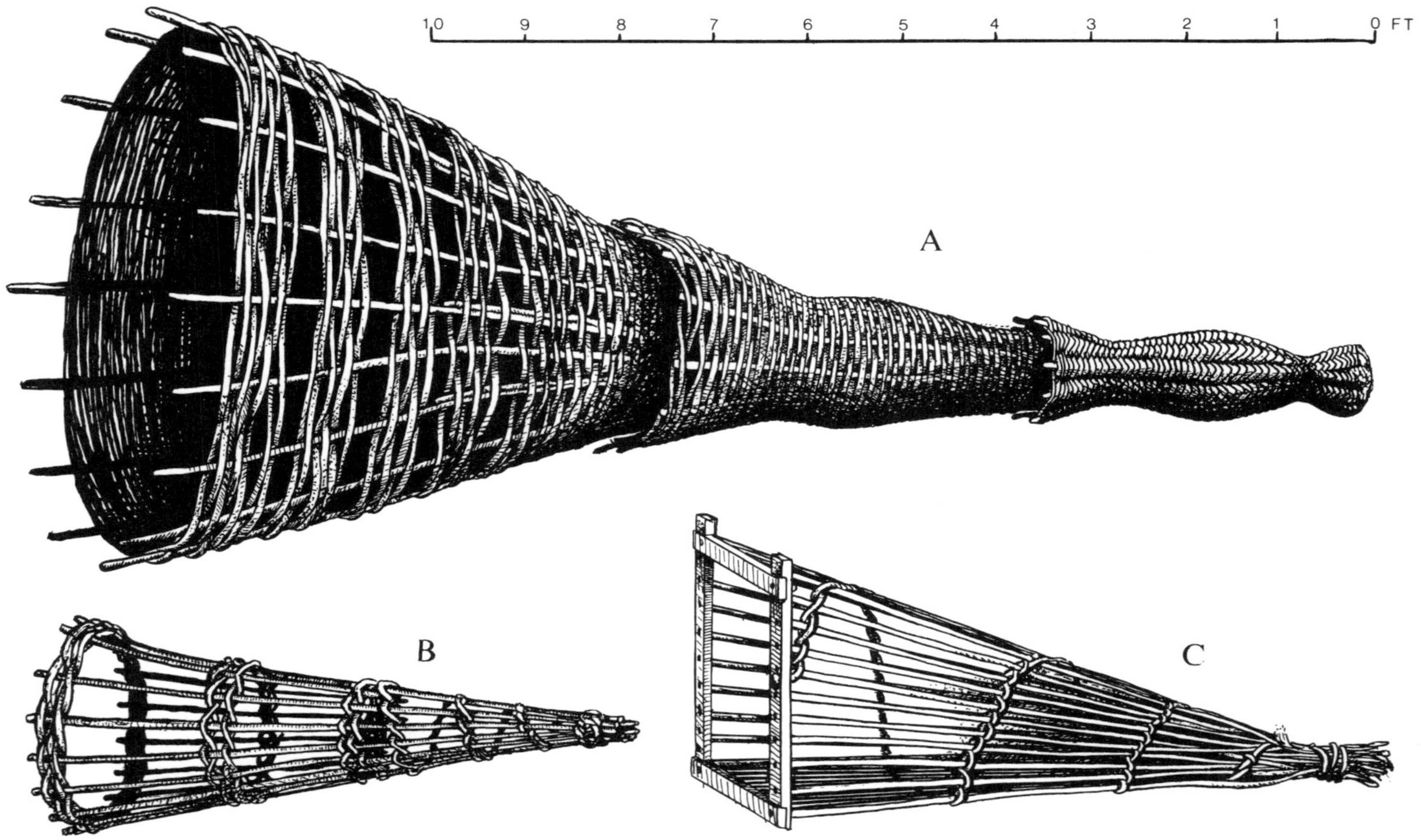

Basket traps: A Putt — Severn estuary; B Putcher — Severn estuary; C Basket trap — River Lledr, North Wales

two inch by one inch timber, measuring thirty four inches long by twenty-seven-and-a-half inches wide forming the mouth of the trap. Thirty-eight uncleft and green hazel rods are inserted in the holes bored in the rectangular frame, each one three-and-a-half inches from the next. The holes are bored at a slight angle to ensure that the rods forming the main body of the trap taper inwards to form a cone. The whole trap measures seventy-four inches in height, and since single rods are not long enough to complete the trap more rods have to be inserted into the body of the basket. A rope is attached to the frame and is twisted in a zig-zag manner around the rods to be tied securely around the tip of the basket. The Lledr trap is now a unique fishing instrument and is kept and used by the occupant of a small-holding, Tan'rallt, who has the sole right of fishing at Pwll-glas on the Lledr.

PUTCHERS

The method of using these entails the construction of a stout timber framework built across the main tidal flow of a river, to carry row upon row of the cone-shaped putchers designed to take fish on the ebb or flow of the tide. 'The salmon, when travelling towards the putcher, passes through the mouth and becomes jammed by the head and is soon drowned by the flow of water past its gills.'[4]

Undoubtedly, putcher weirs are regarded as a most efficient means of salmon fishing, and the high rise and fall of the tide in the Severn estuary is an important factor in allowing a profitable number of baskets to be submerged along workable lengths. The turbidity of the water in the Severn estuary and the Bristol Channel is another factor that has contributed both to the efficiency of basket traps and to their continued use. The weirs may extend into the rivers for considerable distances. At Count Rocks in Gloucestershire, for example, says the *Report of the Inspectors of Salmon Fishing* in 1862,[5] ranks of putts and putchers combined 'aided by hedges of wattles stretch out for nearly three quarters of a mile completely across that moiety of the channel . . . so that if the wind sets to the shore, nearly every descending fish must be captured'. Another observer said:[6] 'The estuary was fished by "putts" and "putchers", each shore being thickly studded with these engines, which were easily erected and self-acting, also unlicensed and worked with a total disregard of all close times; enormous was the damage they did.'

The *First Annual Report of the Inspectors of Salmon Fisheries* published in 1862 provides considerable detail of putcher fishing in the Bristol Channel at that time. The putcher weirs were widely distributed and their presence 'acted as a serious obstacle to improvement and a discouragement to individual exertion'. The Inspectors reckoned that the supply of fish in the Wye, Usk and Severn suffered severely due to the presence of the putcher weirs. Despite the passing of the Salmon Fisheries Act of 1861, which limited fishing 'to those lawfully exercised at the time of the passing of the Act', the number of

Building a putcher rank near Aust Cliff, Severn estuary, 1965

putchers actually increased between 1861 and 1863. The 1861 Report[7] considered putchers as 'specially objectionable modes of fishing, both as being precarious and uncertain in their operation, and injuring the fish for the market by their mode of catching them . . . nothing that once enters them can be taken out unhurt'. One witness at the Enquiry proposed that 'seines and long trammels' should be substituted for fixed engines, for not only were the putchers 'a great obstruction to navigation', but as far as the putchers on the Monmouthshire coast were concerned, fish were prevented from entering the Wye. Later, in 1902, the location of putcher weirs at Porton, Goldcliff and Redwick, between the estuaries of the Usk and Wye, was severely criticised by the Wye Board of Conservators as damaging the salmon catches in the Wye.[8] In 1861, 'it was estimated that between 8,000 and 9,000 putts and putchers were at work. By 1863 they had increased in number to 11,200'.[9] 'The men have been anxious to extend and add to the ranks, thus increasing the number of fixed engines', says the *Second Annual Report of the Inspectors of Salmon Fisheries*. 'In three several cases, summonses were issued against persons for setting up such new ranks on the ground that they had not been exercised at the time of the passing of the Act' (of 1861); 'and in each case, we regret very much to say, the complaint was dismissed, on the ground that as the ancient mode had been exercised at the time of the passing of the Act within the fishery in question, it might be multiplied to any extent. We hope', continues the Report, 'that the first man who sets up a new rank this season will be prosecuted, and if need be that the case may be taken to a higher court. It is one of the greatest importance to the Severn and the other rivers running into the Bristol Channel; for if the decision of the magistrates be correct, there is no limit set by the Act to the number of fixed engines that may be erected; they will increase as the salmon increase; a monopoly will be again established in the estuary, with the inevitable result of neglect of the spawning beds by the upper men, and the consequent destruction of the fisheries. But we believe that decision to be erroneous. It was the intention of Parliament, when it declares that no fixed engines should be used for the capture of salmon, that that prohibition should not extend to such ancient rights or modes of fishing as were then lawfully exercised, but that such engines as were actually in use at the time of the passing of the Act, and *no others*, should be exempted from the general prohibition. Such we believe was the intention, and it seems to us that that intention was most clearly expressed.'

Due to the inadequacy of the 1861 Act in relation to the number of putchers in the rivers and the complete disregard of the weekly closed period between noon on Saturday and 6.00am on the following Monday, the Salmon Fishery Act of 1865 was far more stringent. The Act provided against further increase, titles were closely scrutinised and putcher weirs were strictly limited to those that were in existence in 1861. Those that could not produce charters or show usage from 'time immemorial' were abolished and certificates of legality were issued by the Special Commissioners for English Fisheries.[10] In Table 6 *(see page 54)* the location, ownership and size of putcher weirs is given.

There has been a gradual decline in the number of putcher weirs in the Bristol Channel and the Severn estuary, although the Salmon and Freshwater Fisheries Act of 1923 did no more than reiterate the conditions of the 1865 Act, as far as fixed engines were concerned. Some ranks, like those at Redwick in Monmouthshire and at the mouth of the Rhymney river near Cardiff, just fell out of use for no apparent reason, while others in more recent years were put out of action 'by the Severn road bridge and the structures required to provide cooling water for nuclear power stations on the Severn'.[11]

As far as Wales is concerned, there are only two ranks of putchers remaining in use at the present time – at Goldcliff and at Porton in Monmouthshire, both ranks coming under the jurisdiction of the Usk River Authority. Until 1914, a number of other ranks, at the mouth of the Rhymney, at Redwick and at Undy in Monmouthshire, were in constant use. At the mouth of the Wye near the present road bridge a rank was located at Lyde Rock. This was abandoned in the late nineteen-twenties,

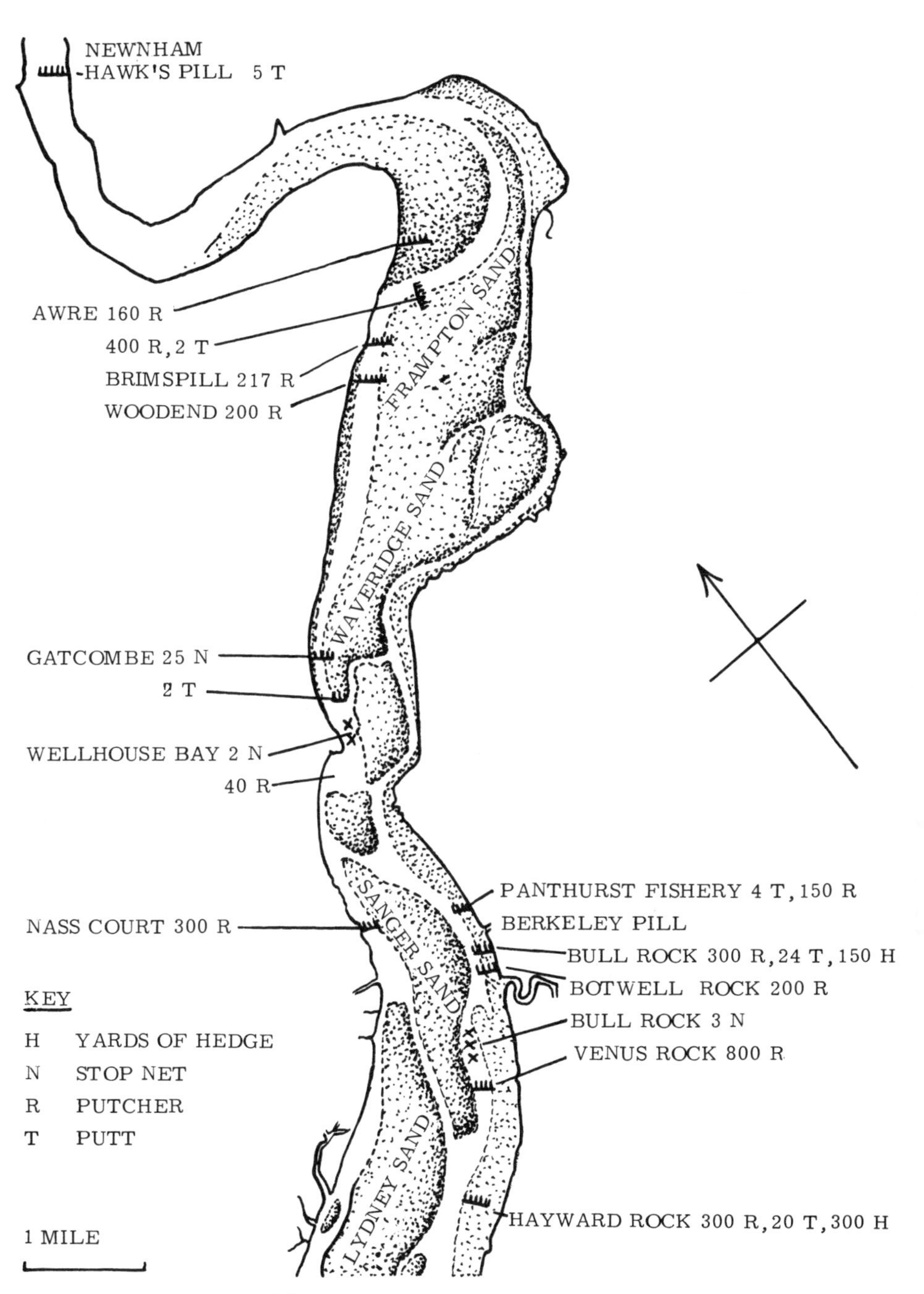

Fishing in the Severn estuary in 1865

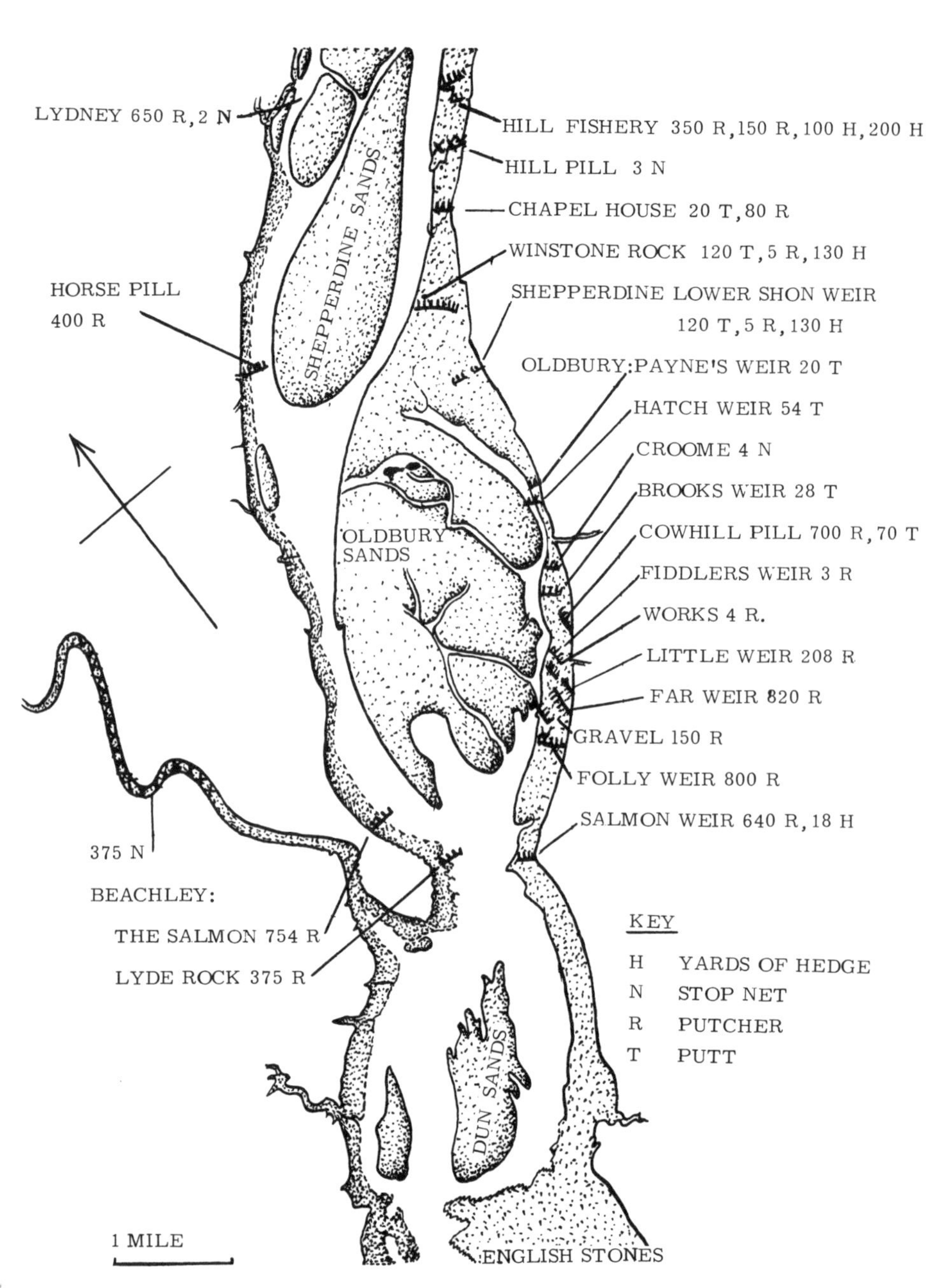

Fishing in the Severn estuary in 1865

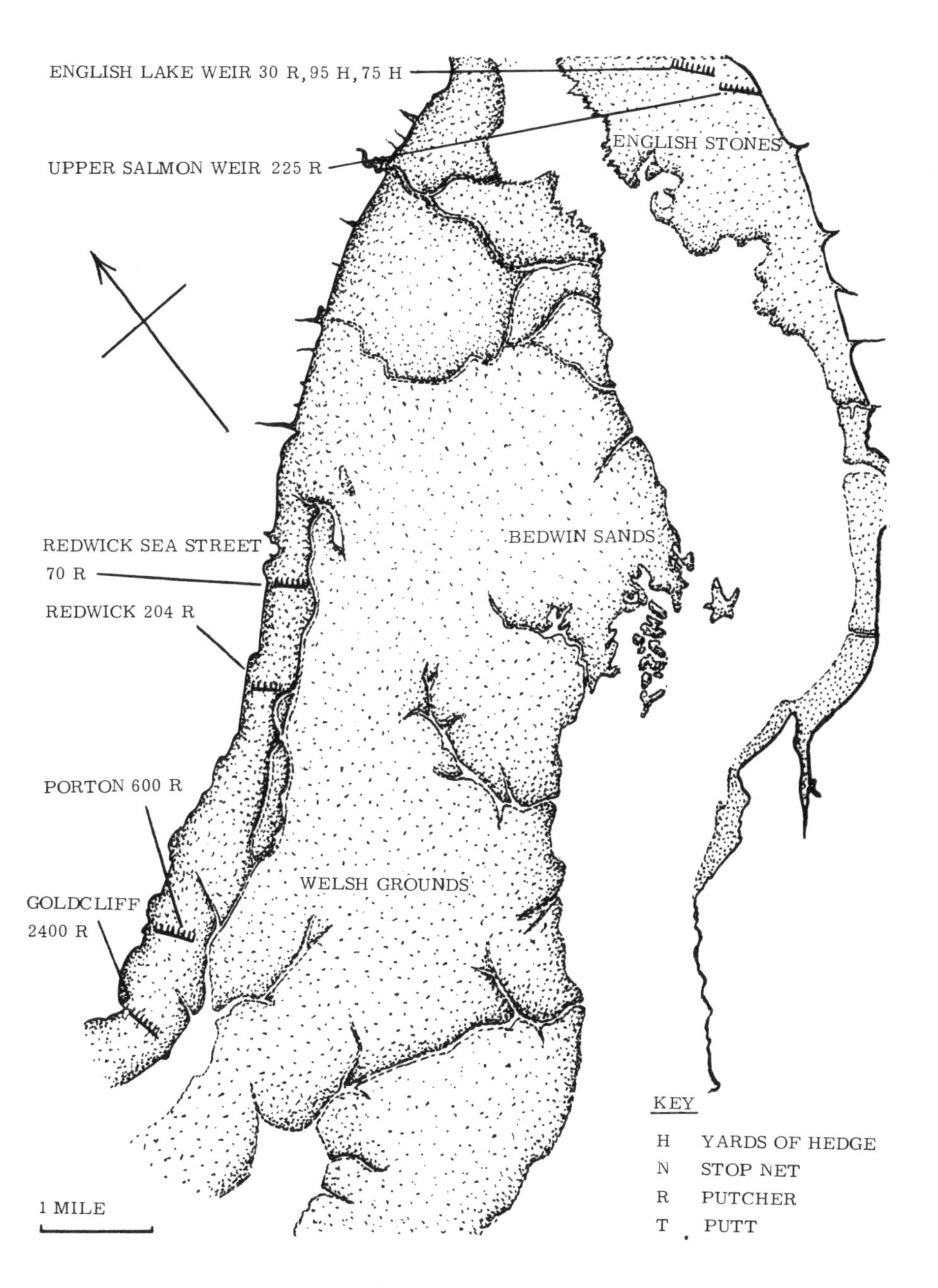

Fishing in the Severn estuary in 1865

TABLE 6
LICENSED PUTCHER WEIRS IN THE BRISTOL CHANNEL 1865-70

RANK	NUMBER OF PUTCHERS	OWNERS	LICENCE
Porton	600	Rev G. Gore and others (Lords of Porton)	1869
Newnham	160	E. W. A. Vaughan	1868
Beachley	754 ('The Salmon')	Duke of Beaufort	1866
	375 (Lyde Rock)		
Shepperdine	80 (Chapel House)	T. C. Barrow	1866
Lydney Pill	650	Rev W. H. Bathurst	1866
Sharpness	800 (Venus Rock)	Lord Fitzharding	1866
	300 (Hayward's Rock)		
	200 (Botwell Rock)		
	300 (Bull Rock)		
Horse Pill	400	Duke of Beaufort	1866
Awre (Brimspill)	400	Thomas Cadogan	1868
Winstone Rock	170	John Cornock	1866
Hill Pill	500	H. J. Fust	1866
Littleton Pill	700	R. S, Holford	1866
Lydney	300	E. O. Jones	1866
Berkeley Pill	150	W. J. Phelps	1866
Aust	150 (Gravel Weir)	R. C. Lippincott	1866
	800 (Folly Weir)		
	640 (Salmon Weir)		
Redwick (Glos)	225 (Lower Salmon Weir)	R. C. Lippincott	1866
	300 (Upper Salmon Weir)		
	30 (English Lakes)		
Redwick (Mon)	70 (Sea Street)	Lords of Porton	1869
Awre	204 (Redwick)	Trustees of Sir Thomas Rice Hospital, Gloucester	1866
	217 (Woodend)		
	200 (Brimspill)		
Oldbury	54 (Hatch's Weir)	Aaron Taylor	1866
	3 (Fiddler's Weir)		
	4 (Works Weir)		
	208 (Little Weir)		
	520 (Far Weir)		
Wellhouse Bay	40	Walter Taylor	1866
Hill Pill	400	H. J. Fust	1866
Oldbury Pill	50	William Trotman	1866
Goldcliff	2,400	Eton College	1866
TOTAL	13,354		

and the Beachley putcher fishery at Slime Bay, administered and maintained by employees of the Wye River Authority, now carries a single rank of 600 putchers about half a mile up-river from the Severn Bridge. The Horse Pill putchers further up-stream, although owned by the Wye River Authority, are let to a private tenant. The Severn River Authority operates half a dozen weirs on both sides of the river accounting for a total of 4,047 putchers in 1970, while in Somerset, a putcher weir was in use in recent years near the mouth of the River Parrett.

'The baskets', as they are termed by the fishermen themselves, are made of willow or wire. Willow putchers are used exclusively by the Wye River Authority at Beachley, but at Porton and Goldcliff the all-willow putcher has been gradually superseded by ones of metal. Initially, galvanised wire baskets were adopted by the fishermen from 1942, but from 1952 baskets made from sea-resistant aluminium wire have been used at Goldcliff and Porton. A few of the older galvanised wire putchers are still used at Porton, but in 1970 the Goldcliff fishermen made and installed about a dozen of the traditional baskets amongst the three ranks that carried a total of well over two thousand putchers, most of them being of aluminium wire. Although metal has replaced willow as the most common material for making putchers, traditional measurements are still adhered to by the manufacturers,[12] each putcher being from five feet to six feet long and two feet three inches in diameter at the mouth, tapering to a diameter of five-and-a-half inches to six inches at the nose. Galvanised wire putchers of three inch square welded mesh were supplied flat and had to be wired into shape by the fisherman. The aluminium putchers with diamond mesh, costing from £2 to £3 each, are used today at Porton, exactly as supplied by the manufacturer, but at Goldcliff each putcher is strengthened by the addition of wire bands. To fix metal putchers in the fishing position on the wooden framework of the weir, two staples at the front and one at the rear are driven into the rails on which each putcher rests.

To make the older willow putcher, autumn-cut withies,

Salmon basket or 'frail' used for carrying fish from the putcher rank at Beachley, Severn estuary, 1971

usually from a six acre plantation at nearby Llanwern, are used. The baskets are made by the fishermen themselves during the close season between 14 August and 1 May. The cutting of withies and putcher making, together with repairing the framework of the putcher ranks, provides the winter's work for the fisherman. To make a putcher, a low bench eighteen inches high and twenty-four inches square is required. A twelve inch ring with nine round holes is drilled in the top of the bench and this stands over a narrow, conical pit, twelve inches deep. Nine withy rods are passed through the holes in the bench, and a withy ring, approximately ten inches in diameter, is plaited around the rods, close to the surface of the bench. Although withy rods 'in the round' are occasionally used by Severn-side putcher makers, it is far more common to split each rod into three sections, using an oak cleaver held in the maker's palm. With the withy plait firmly in place nine shorter rods, either cleft or in the round, are then inserted in the ring and another ring is plaited about half way up the rods and another at the tip. The ring at the nose is plaited and then coiled back in between the rods to be attached to the lowest plaited ring on the bench. The putcher is then removed from the work bench and the base ring, forming the mouth of the putcher, approximately twenty-four inches in diameter, is then plaited in a similar manner. Withy baskets are expected to last for two fishing seasons, and possibly a third season after repair. The method of using putchers entails the construction of a stout timber framework, built across the main tidal flow of a river.

At the Goldcliff fishery there are three ranks – 'flood', 'ebb' and 'putt', carrying a total of 2,327 putchers; each rank is from two hundred feet to three hundred feet in length. The Priory of de Chandos, its lands and fishing, were given to Eton College on its foundation during the reign of Henry VI[13] and the right of fishing between Goldcliff and the old Porton Estate was owned by the College and leased to a fisherman. Until 1919 when the College disposed of the fishery, the Goldcliff weir consisted of 2,400 baskets and the revenue from the property provided an endowment to the Church of the Blessed Virgin at Eton. In the past 'scholars of the college were granted allowances for breakfast from the catches of the Goldcliff fishery'.[14] The present owner of the fishery is Mr John Williams, who once owned the Beachley-Aust ferry, and he employs one full-time putcher fisherman.[15] When the fishery was owned by a Newport fishmonger in the nineteen-twenties, three full-time putcher fishermen were employed at Goldcliff.

The Porton fishery, about a mile away from Goldcliff, carries a single rank of 600 putchers and was acquired by the Pontypool Park Estate and subsequently leased to various tenants. It is now tended in season by the occupant of a nearby farm and managed on the owner's behalf by the estate surveyor's office. No one knows when putcher fishing began on this section of river, but the Manorial Survey of 1663 mentions fixed engines between the 'Hill and the Pill'.[16]

The putcher ranks at Porton and Goldcliff are located on an even slope, and the weirs are built on a framework of green larch or elm poles, up to fifteen feet long and eight inches or nine inches in diameter at the butt, sunk to a depth of from six feet to eight feet. No treatment of the timber is considered necessary, for the salt in the water preserves the timber for ten years or longer, when they begin to rot from the top. It is said that some of the elm poles in the Goldcliff ranks have been in place for over forty years, and elm is regarded as being far more durable than larch for constructing putcher ranks.

To bore the post holes a rock auger and bar are used, and the debris is removed from the hole with a long-handled ladle, known as a 'spoon'. The uprights are beaten into place with an iron-lined yew beetle, in pairs five feet apart to make a double row with a six foot spacing between each pair. Each pair is always braced transversely or diagonally near the top, and sometimes lower down, as well. Nailed to the outside and running the full length of the rows are four or five lines of horizontal rails, one above the other, two feet apart, the lowest eighteen inches clear of the ground. Two or three narrow gaps are left at intervals in the lower rails as throughways to facilitate collection of fish and maintenance. Rows of baskets lie across the frame, tiered one above another, supported at the front and rear by opposite rails, which are so positioned that the baskets are tilted upwards at an angle of approximately twenty degrees; this is in order that trapped fish may be pressed downwards by force of the tide rather than be washed back out. The mouths of the baskets all face to the same side, referred to as the 'front' of the rank, which may be toward either the incoming or the outgoing tide. At Porton and on the putt rank at Goldcliff, outlying props slope from the leading post of about every fourth pair and are anchored to short staves driven in a few feet to the rear, the front of the rank being kept clear. At Porton, where the mud has deepened, a walkaway runs full length down both sides, just below the bottom baskets, in some places no wider than a single plank. Although during this century both a lighter and a tug boat have been driven onto the 'Flood' rank, only in 1968 were marker poles made compulsory at both ends of each fishing, not less than five feet to be visible at high tide, and carrying a luminous panel at least 18in by 12in. The 'Flood' rank now has a forty foot pole held by three chains and capped by a milkchurn wrapped in fluorescent paper!

Until the early years of the nineteenth century, putcher entrances had to be covered from midday on Saturday until 6.00am on the following Monday. Nowadays, fishing can continue without ceasing throughout the season, and day and night immediately after the recession of the tide the ranks have to be inspected. Armed with a pronged staff, the fisherman walks along the rank, pushing any fish caught in the basket back towards the entrance. On the return journey along the rank the salmon are packed in a wet sack. At Goldcliff the fisherman wheels a bicycle equipped with a tray at the front and back to carry the fish. A bag of nails and a hammer are also essential for repair work on the ranks. The larger ranks at Goldcliff, the putt and flood, are so sited that there is ample time to inspect the ebb rank before the lower ranks are exposed. The fisherman's brick workshop is equipped with a bed and cooking equipment, for the putchers have to be inspected in the night as well as in the daytime.

The fish house at Goldcliff is a lofty single-storied building with wide slatted vents high up on the pine ends. Fish are placed in a tight-lidded, lead-lined wooden chest with broken ice. The salmon are then sent to Billingsgate. At Porton fish are taken to a cool farm outhouse and placed immediately in Billingsgate boxes, ready for transport to that market.

An interesting building at the Porton fishery is a brick smoke house eight feet square and eighteen feet high with a steeply sloping roof of sheet tin with vents near the top. This was built some thirty years ago for smoking salmon. Fish were suspended high up on wires and oak shavings were kept smouldering on the floor below. This venture was short lived and the smoke house has been unused for many years.

PUTTS

Putt fishing, using large wicker baskets moored in a line on the river bed, was never as widespread in the Bristol Channel as putcher fishing. It has completely disappeared from the rivers of south-east Wales, the last putt weirs at Goldcliff with 100 baskets and at the Rock Fishery of 127 baskets between Porton and Goldcliff being abandoned in the nineteen-twenties.[17] The baskets for these weirs were made by a basket-maker at Redwick in Monmouthshire, and they were designed for catching all types of fish, ranging from salmon to shrimps. Indeed, according to the Goldcliff putcher fisherman the main function of the putt weir was to catch shrimps which were actually boiled at the fishery, although large quantities of salmon, flat fish and even the occasional sturgeon were caught in the putts.

It has been suggested[18] that putt fishing is a considerably older method of fish capture than putcher fishing, for 'Reference is made to them in 18 Geo. 3 c. 33 as "the ancient putts", to which at some places an inside wheel or diddle is fixed, whereby great quantities of the spawn or fry are taken and destroyed, and we are led to believe that the putcher . . . is of modern invention: an improvement in short upon the ancient putt.'

One of the main reasons why putts have largely fallen out of

A putt with kipe, butt and fore-wheel at Awre, Severn estuary, 1965

use in south-east Wales is the fact that they demanded far greater skill in the art of basketry than the more simply constructed putcher. The latter with its open weave can be constructed by a comparatively unskilled worker, but the putt with its three sections fitting firmly into one another demanded much more skill. Both butt and fore-wheel are urn-shaped with distinct bulges in the centre, while the cone-shaped kipe not only demanded skill in its construction but also considerable strength in randing stout withy, hazel or whitethorn rods.[19] The ribs of a kipe may consist of a dozen or more sets of sticks, each set consisting of five or six hazel sticks placed close together to give the kipe added strength. Withy is then woven in between the ribs, the withy becoming less stout and smaller in diameter towards the nose of the kipe. Usually the ribs on the butt consist of three or four sticks placed close together and running the whole length of the butt, the withy being less stout than that on the kipe. In the fore-wheel only single rods are usually used for the ribs, while the osiers used in weaving are even finer than that used in the kipe. Some of the putts used in recent years on the left bank of the Severn estuary are less carefully constructed than the traditional putts, and the kipes used in a rank at Oldbury are made in the manner of a putcher, woven loosely with hazel and willow. The butts too, although closely woven, are cone-shaped rather than urn-shaped.

The three sections of the putt are kept firmly in place on the river bed by a series of wooden pins placed on either side of kipe and butt with withy loops known as 'rods' passed over the kipe and butt. 'The rod is a thong of withy, expertly twisted, for in the hands of a basket fisherman a two year old shoot of withy can be turned into a stiff rope in a few seconds. The growing end of the withy is held under the foot while the fisherman spins the wand with a rapid movement of his hands; the withy thus pliant is twisted back on itself and plaited like rope in tight spirals to a rod, leaving a noose smaller than a penny at the end. The rod is bound through the kipe on to the stake, but no tie is made. The pin, a stout six inch peg of hazel, is pushed through the noose of the rod, a sharp twist like a tourniquet is then made and the pin is driven home.'[20]

The fore-wheel is kept in position with a forked stick driven firmly into the ground. A withy loop known as an 'apse' encircles the fork and the fore-wheel. Both the construction of putts and their positioning in a weir demand considerable skill, and this may be the reason why putts have virtually disappeared from the Severn estuary in recent years, despite the fact that 'putts are the most efficient method ever devised of fishing in the estuary'.[21] In putt fishing, the catch may be removed by taking out a bung of wood or kelp and emptying the contents into a shoulder basket, known as a 'welch' or 'witcher'. The process of unhasping the fore-wheel before tipping the catch in a basket is known locally as 'cunning putts'. 'A kipe is so cumbersome and heavy', says one author,[22] 'that it is usually dragged out to its mooring on a sled. When fixed to a weir it remains there until lost, destroyed or worn out, but at the end of the salmon season in August the butts and forewheels are removed, allowing fish to pass through the kipe without danger.'

Like the lave-net fishermen, the putt fishermen of the Severn have to take great care in attending their weirs. 'The great thing', said one writer,[23] 'is to allow yourself more than adequate time to get ashore before the tide starts running, for the Severn tide is notorious for its rapidity and power. One of the fishermen told me that he always carries two watches with him, when he

A putt rank at Shepperdine, Severn estuary, 1972

goes out alone. He was once nearly drowned because his watch had stopped and he had miscalculated the time.'

In the past it was customary to build brush or wicker-work hedges on one or both sides of a putt weir so as to drive or guide the salmon on to the front, fishing face of the weir. This method of hedging was also commonplace with putcher fisheries, especially on those located near the left bank of the Severn. In some cases such as at the Bull Rock Fishery putts and putchers were mixed, so that there were two ranks of 300 putchers, with a single rank of 24 putts between the shore and the putchers and a line of wickerwork hedging at right angles to the weir, to ensure that the fish were carried towards the weirs on the ebbing tide.

In 1970 the total number of putts licensed by the Severn River Authority was eight with three leaders, most being in the Oldbury district on the left bank of the river.

THE TAN'RALLT BASKET TRAP

The River Lledr, a tributary of the Conway, is about eleven miles long and is considered to be an excellent salmon river. The basket trap at Pwll-glas on that river authorised by the special commissioners in 1867 on the grounds of 'immemorial usage' is the only trap of this kind left in Wales. The commissioners specified the size of trap and the exact position in the river where it was to be used. They stated that it had 'to be placed at a natural fall in the channel on the right bank of the River Lledr. On the right hand of the river there is a portion of the rock which is generally quite dry, but when the river is in flood the water comes over the rocks and falls down into a pool . . . and from the pool the only exit is by means of this narrow channel lying between the rocks'.[24] The basket trap is placed in this channel about Pwll-glas and is especially shaped to fit tightly into a natural cleft in the rocks. It is placed in a fixed wooden frame at a point where the channel begins and is fastened to the rocks by staples and chain. It is only when the river is in fair flood that salmon are likely to pass through the channel, and even then only what are termed 'the flood fish' take that course instead of swimming in the main stream. Nevertheless appreciable quantities of salmon have been caught in this single basket; the highest number, it is said, being a total of 281 fish weighing 1,942 pounds in 1888.[25]

The trap remains in the fishing position in the river until the flood has subsided enough to expose at least a part of the trap. 'The fishermen', said a witness at the Appeal Court in 1887,[26] 'come down the steep rocks and fasten a rod to the basket and lift it up to the top of the rock and empty its contents, but if they were to do that without putting something else down in the meantime, any fish that might fall out of the first basket or might remain in the pool, would have an opportunity of getting away, and therefore a similar basket is put immediately behind that point and lies there when the first one is being hauled up

The Tan'rallt fish trap, used on the River Lledr, a tributary of the Conway

and put back again. When the first one is lowered again, the other is taken up.' The fact that the fisherman used two baskets, plus a net below the trap and in the pool, led to litigation in the eighteen-eighties, for the commissioners only allowed the use of a single basket at Tan'rallt. As a result of the alleged contravention of the conditions of its usage[27] John Jones, the licensee, was fined £10 or imprisoned for two months in default of payment. The baskets at that time were '7 feet high outside, 22 inches at the mouth and 26 the other way. The distance between the hazel rods at the mouth was 2 to 2¼ inches. They came close together at the bottom'. At that time the tenant of Tan'rallt, John Jones, held two licences – one for the salmon trap and the other for a coracle net on the Conway.

There was further legal action in the nineteen-twenties due to 'the alleged infringement of a bye-law which forbids the taking of fish by any other means than rod and line on the Sunday'.[28] The Tan'rallt basket trap, it was stated, 'is an unique fixed engine for the taking of salmon, which has . . . been in the possession of the one family for over 300 years and which the present owner – a lady over 80 years of age – is anxious to preserve'. At the quarterly meeting of the Conway Fishery Board, the owner maintained that 'to compel her to have the cage or basket lifted out of its place every Saturday night and replaced every Monday would give needless trouble and would deprive her of immemorial rights'. The case at that time was dismissed, but since then it has been obligatory to discontinue the use of the trap on a Sunday in accordance with the Salmon Fishery Act of 1923.[29] A wooden frame carrying wire netting is placed over the entrance to the trap on a Saturday and must not be removed until the following Monday.

For notes to Chapter 3, see pages 307-8

4 Net-Making

Today, most of the nets used in England and Wales are manufactured by a large-scale Bridport manufacturer who is able to produce the variety of netting that local custom amongst fishermen has dictated should be used in the various rivers. The manufacturers can manufacture nets that fulfil conditions of design, size and mesh specified by the river authorities. Although some fishermen may still braid their own nets from bast fibres such as hemp and flax, the fishing industry has become heavily dependent on commercially manufactured nets, most of which are made from synthetic fibres such as nylon, terylene and polythene.

All the nets purchased from Bridport are machine-made nets produced on a complex multi-shuttled netting loom, with its arrangement of hooks, needles and sinkers that form the meshes of a net. From the twelfth century, Bridport has been the all-important centre of rope- and net-making in Britain. Undoubtedly one of the main reasons why Bridport became pre-eminent in the manufacture of cordage was that the soil and climate of Dorset were particularly suitable for the growth of flax and hemp and the early industry was stimulated by the demand of fishermen and boatbuilders who flourished on the Dorset coast. By the end of the sixteenth century Bridport, with its superlative hemp and flax and with its long start over other centres of production, had achieved a virtual monopoly of rope and net production in Britain. Although there was a decline in the industry in the early eighteenth century, the second half of the century saw the return of prosperity, with nets of all kinds, rather than rope, becoming the most important product. At the turn of the nineteenth century, a contemporary writer[1] noted that 'the manufacture of Bridport perhaps flourishes more than at any other former time and furnishes employment not only for the inhabitants of the town, but for those likewise of the neighbouring villages to the extent of ten miles in circumference. It consists of seines and nets of all sorts, lines, twines and similar cordage and sail cloth.' As in the wool textile trade in Yorkshire, Wales and elsewhere, hemp merchants who organised the manufacturing processes were common in the Bridport district. Samuel Grundy, for example, was a merchant and possibly a banker who, as early as 1665, purchased the hemp crop and issued it to families for conversion into yarn and nets in their homes, and then marketed the goods. Until well on into the nineteenth century, most of the net-making was carried on as outwork in the cottages of Bridport and the surrounding countryside, and methods of braiding by hand were handed down the generations from mother to daughter. Net-making was organised from a central factory; twine was delivered weekly and the finished nets were collected when next week's work was delivered. John Claridge estimated that in 1793[2] there were 1,800 people concerned in the rope and netting trade in Bridport and a further 7,000 in the surrounding countryside.

During the nineteenth century imported hemp replaced the locally grown crop for the manufacture of nets and the century, too, witnessed the growth of a factory system in the Bridport industry, with new machines being introduced to spin and hackle fibres. Hand looms, the early ones known as 'jumper looms', became commonplace after 1860. The jumper loom, invented in the eighteen-thirties, was so called because of a wooden pedal on which the operator had to jump. This formed the knots around three rows of hooks and drew twine for the next meshes by means of a long needle-like, barb-tipped shuttle. Thrifty families were able to purchase one and with the purchase of a second loom and hiring an operator many an individual's cottage workshop developed into a small factory. By 1900 power looms were being introduced into the industry and factory production of nets was in the hands of fifteen family firms. Today, the industry that supplies most of the netting requirements of the fishing industry is in the hands of one large-scale manufacturer, although a certain amount of domestic manufacture is still practised. 'Despite looms which tie knots at unbelievable rates, the net making industry still has to fall back on outworking skills which are based on pure pre-Industrial Revolution cottage industry, and has to dispatch vans daily to take out the raw material, collect the finished nets and pay for them.'[3]

Traditionally, hemp twine was preferred by all fishermen for fishing nets, but between the two World Wars low-priced cotton was widely used. It was regarded as being much lighter than hemp, despite its inferior strength. Since 1945 synthetic fibres have become increasingly more important in the netting industry although for some nets, such as coracle nets and Dee trammel nets, fishermen still insist on traditional materials. Nets of nylon, terylene, polyethylene and polypropylene have, according to some, many disadvantages. They are affected by sunlight; some create static electricity and many, being made of very thin twine, can cut off the heads of fish. It is not surprising, therefore, that some fishermen still braid their own nets from hemp and cotton and equip them with headlines of horse- and cow-hair, much in the manner of past centuries.

Nets used by fishermen today have to fulfil conditions of length, depth, mesh and design specified by the regulations of the various river authorities. These can vary tremendously, as is shown in Table 7 *(see page 70)*, giving net sizes from current issues of fishery bye-laws.

The process of making a net by hand is known as 'braiding' or 'beating' and though today most of the nets used by commercial fishermen in England and Wales are factory made from synthetic materials, many fishermen, nevertheless, still prefer to braid their own. Ordinary, simple rectangular pieces of netting, such as are found in drift nets and seine nets, can be made by machinery, but most of the nets that have to be braided to a special shape have to be made by hand, although shaped pieces can be cut from machine-made net and laced together. The process of hand braiding demands very little in

TABLE 7
TYPES OF NETS USED FOR THE CAPTURE OF SALMON AND MIGRATORY TROUT IN FISHERY DISTRICTS IN ENGLAND & WALES, 1972

RIVER AUTHORITY	TYPES OF NET	MESH (KNOT TO KNOT)	DEPTH	LENGTH
Wye	Drift (Tuck)	2¼in	10ft	Not specified
Severn	Seine (Draft)	Not specified	Not specified	200yds (below Lower Parking) 100yds (elsewhere)
	Lave	Not specified	–	–
Usk	Drift (Tuck Net)	2in	9ft	300yds
	Lave	2¼in	–	7ft 6in between staffs
Dee & Clwyd	Seine	2in	5yds	200yds
	Trammel	2¼in & 11in	2yds	100yds
	Drift (Sling Net)	2in	5yds	200yds
Gwynedd	Seine	Not specified	6yds	150yds
South-West Wales	Seine (Tywi)	1½in	Not specified	200yds
	Seine (Teifi)	2in	Not specified	200yds
	Coracle (Tywi)	1½in & 5½in	3ft 9in	33ft
	Coracle (Teifi)	2in & 5½in	3ft 9in	20ft
	Wade Net	2in	6ft	30yds
Devon	Seine	Not specified	8yds	200yds 280yds (Lower Exe)
Cornwall	Seine	Not specified	10yds	200yds
	Drift	Not specified unless trammel is used - 11in mesh	10yds	200yds
Yorkshire Ouse	Drift	Not specified	Not specified	400yds
	Seine	Not specified	Not specified	400yds
Lancashire	Drift (Whammel)	3¼in (Ribble)	3yds	150yds
		2½in (Lune)	3yds	320yds (Lune)
	Haaf	2in	4ft	18ft
	Lave	2½in	–	6ft 6in (max)
	Seine	2in	8yds	200yds

TABLE 7 *(continued)*

RIVER AUTHORITY	TYPES OF NET	MESH (KNOT TO KNOT)	DEPTH	LENGTH
Somerset	Dip	–	–	24sq ft entry
Cumberland	Seine	1in	Not specified	Not specified
	Drift	1in	Not specified	Not specified
	Haaf	1¾in	Not specified	Not specified
	Coracle	Not specified	Not specified	Not specified
Northumbrian	Drift	1½in	Not specified	600yds
	Seine	1½in	Not specified	600yds
	T-Net (Fixed Engine)	4in (Bag Net) 5in (Straight)	–	200yds (Bag) 300yds (Str)

the way of equipment, but requires considerable dexterity in the manipulation of twine. The twine is first of all wound on a special hardwood 'needle', which is a wooden shuttle, flat in cross-section and varying in size according to the size of net being made. The mesh is regulated by forming each one over a piece of wood, oval in cross-section and known as a 'spool'.[4] In braiding a net, the twine is wound on the needle by being passed alternately between the fork on the back of the needle and round the tongue in its centre, so that the turns of the twine lie parallel to the length of the needle and are kept on by the fork and tongue.

The techniques of braiding can vary tremendously, but to make the simplest form of rectangular net, such as a seine or drift net, the process is as follows. A length of line is stretched out taut, usually between two stakes and on to it is set a row of 'bights' or 'loops' of the netting twine. 'These bights are attached to the line by a series of "clove hitches" say from right to left and they form the first "row of half meshes" or "round" or "overing". The second row of half meshes is then made on to the first, from left to right, the knots of the second round being made on to the centres of the bights of the first round, thus completing the first "row of full meshes". After the completion of the second round, the process is then repeated from right to left, and thus the third round is made. The

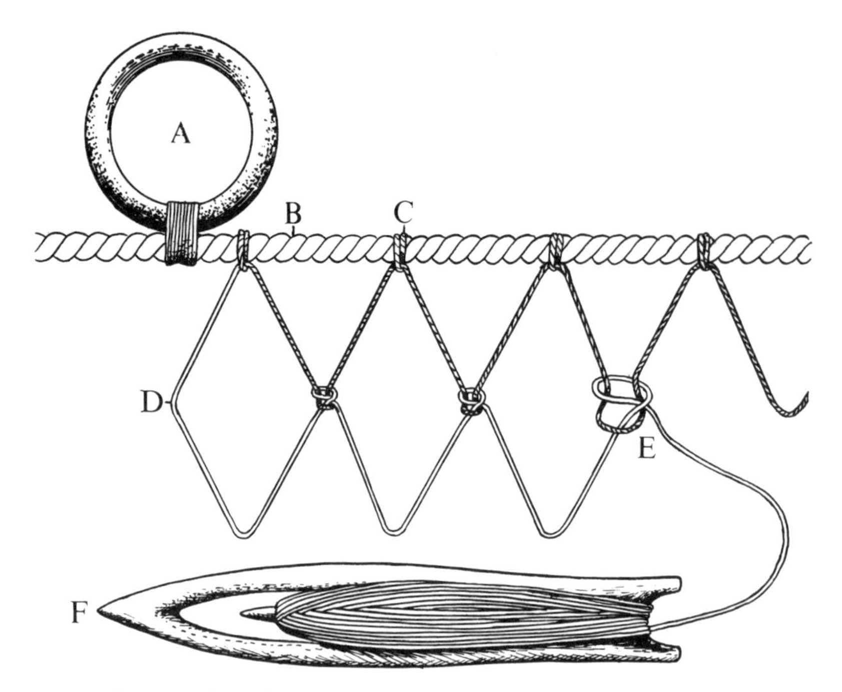

Construction of a hand-made net: A ring; B cork line; C clove-hitch attachment for first row of meshes; D selvedge; E sheet-bend knot; F braiding needle

completion of the third round closes the half meshes remaining between the bottom halves of the full meshes formed by the completion of the second round, so that two interlocking rows of full meshes are now formed, although the depth of the net is only three mesh sides or "bars". This process can be repeated indefinitely until the whole piece of net is formed. It is usual after braiding a few rounds, to remove the original line, thus casting off the clove hitches and leaving free the first row of full meshes, which are then threaded on a line or thin rod.'[5]

This is the basic method of braiding nets, but many variations occur both in technique and in the design of the finished product.

For example, variations occur in methods of making knots, which can be variations of single sheet bends or reef knots. Sheet bends may vary and bear different names such as 'trawler's knot', 'shoot knot', 'beating knot', 'herring knot', 'round knot', 'braiding knot', 'understitch' and 'underbraiding' while net reef knots may bear such names as 'flat knot', 'freshwater knot', 'beating knot' and 'baggars knot'. In some cases, to make a stronger net, double knots are made; indeed with the advent of synthetic fibres, double knotting is essential because those fibres have a smooth, slippery surface and knots can slip. In the net-making industry, double-knotting machines are now common.

In a net, the end at which the work is started is known as 'the head' and that at which it is finished as 'the foot'. The side edges are 'the selvages' and the number of meshes in a 'round' or row can be increased or decreased as braiding goes on. Increasing the width of a net is known as 'creasing' (ie increasing)[6] and decreasing the width as 'bating' (ie abating).[7] The simplest method of creasing the number of meshes in a round is 'on the completion of any mesh to make a small mesh on the same bight as the mesh just finished, before continuing on to the next bight. On returning on the next round, this small mesh, the knot of which is coincident with the knot of the parent mesh, will be treated as a full mesh thus increasing the breadth of the piece of one mesh'.[8] The simplest way of bating is to include two bights of the previous round in the knot to complete a mesh.

Variations occur in methods of selvaging and sloping the edge of a net. This may be done by bating or creasing or it may be done by 'fly meshing'.[9] 'This simply consists of missing out the last mesh on one selvage every time the needle is passing towards it, thus reducing the breadth by one mesh every other round.'[10] Variation can also occur in the shape of net meshes, for although most are diamond shaped, some like the armouring of a trammel net are rectangular. The basic method of braiding is similar as for diamond meshes.

A fishing net is attached to a head-rope[11] and a foot-rope,[12] but a wall of netting, sometimes called a 'lint', 'slint' or 'linnet',

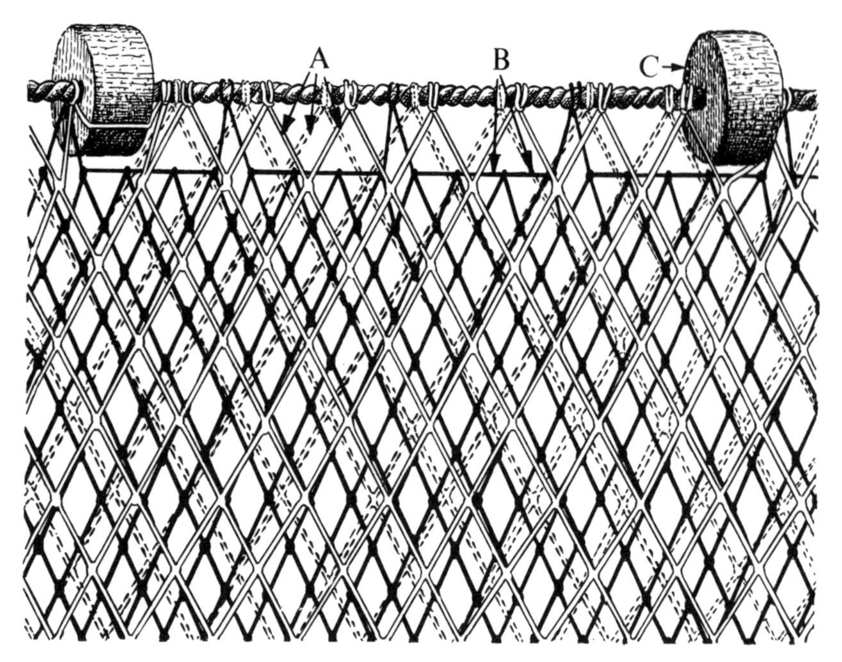

Construction of a Dee trammel net: A Lint; B Armouring; C Cork float

will be considerably wider than the two ropes when fully stretched. For the meshes of a net to hang diamond-wise, a fisherman reckons that three meshes sould occupy the lateral space of two fully stretched meshes. This method is known as 'setting in by the third' and is by far the most common method of setting drift and seine nets. If, as in some drift nets, four meshes occupy the space of three, this is known as 'setting in by the fourth'. In trammel nets, such as the type used on the Dee, the central net or lint, attached to the same lines as the outer nets or armourings in a two- or three-walled net, must have loosely hung meshes. In this case the net is 'set in by the half'; that is the two meshes occupy the space of one. 'If 2¼ or a similar number of meshes occupy the space of two; then the setting is "within third"; if it is 3¼ meshes in the space of two, it is "without the third".'[13]

As far as vertical setting is concerned, nets such as seines have to be set in such a way that the wall of netting forms a distinct bag in use. This is achieved by making the vertical end ropes shorter than the actual depth of net. 'Thus the arms of some seines may be from 12 to 18 feet in depth, but will usually be fixed to end ropes or poles of 6 feet or less in length, thus leaving a certain amount of depth or "flow", which may be increased towards the centre of the net by having an extra depth of lint.'[14]

Davis notes[15] that there are three distinct ways of setting a wall of net. 'First, 90 yards of net may be set in to 60 yards by "reeving" a line 60 yards in length through the head meshes, this method being more or less confined to stake nets and the head-ropes of some beam trawls. In a few nets the setting and meshes are combined by braiding the head and foot meshes direct on to the head-line and foot-rope by means of clove hitches. The diamond armouring of some trammels affords the best example of this.

'The second main method is that of "stapling". Again assume that 90 yards of lint is to be set in to 60 yards and that the meshes are 1½ inch bar, ie 3 inch diagonal. A light line is hitched to one end of the main line, and its free end is passed through three meshes of the head of the net, and then set on to the main line, by a clove hitch or other knot, at a distance of 6 inches from the first knot; so 9 inches (diagonal) of net hangs from 6 inches of head-rope, and the process is repeated all along the line until all the head meshes have been dealt with. Each of the bights of light line thus formed is a "staple". In most cases the staples hang loose as in the figure and are not stretched taut.

'The third method is "norselling" or "osselling". In this case, on to about every sixth knot along the head, a small line is fixed, each being about 10 inches in length and ¼ inch in diameter. These are the "norsels" or "ossells". Assuming that the net is to be set in by the fourth, four of the norsels are fixed by their free

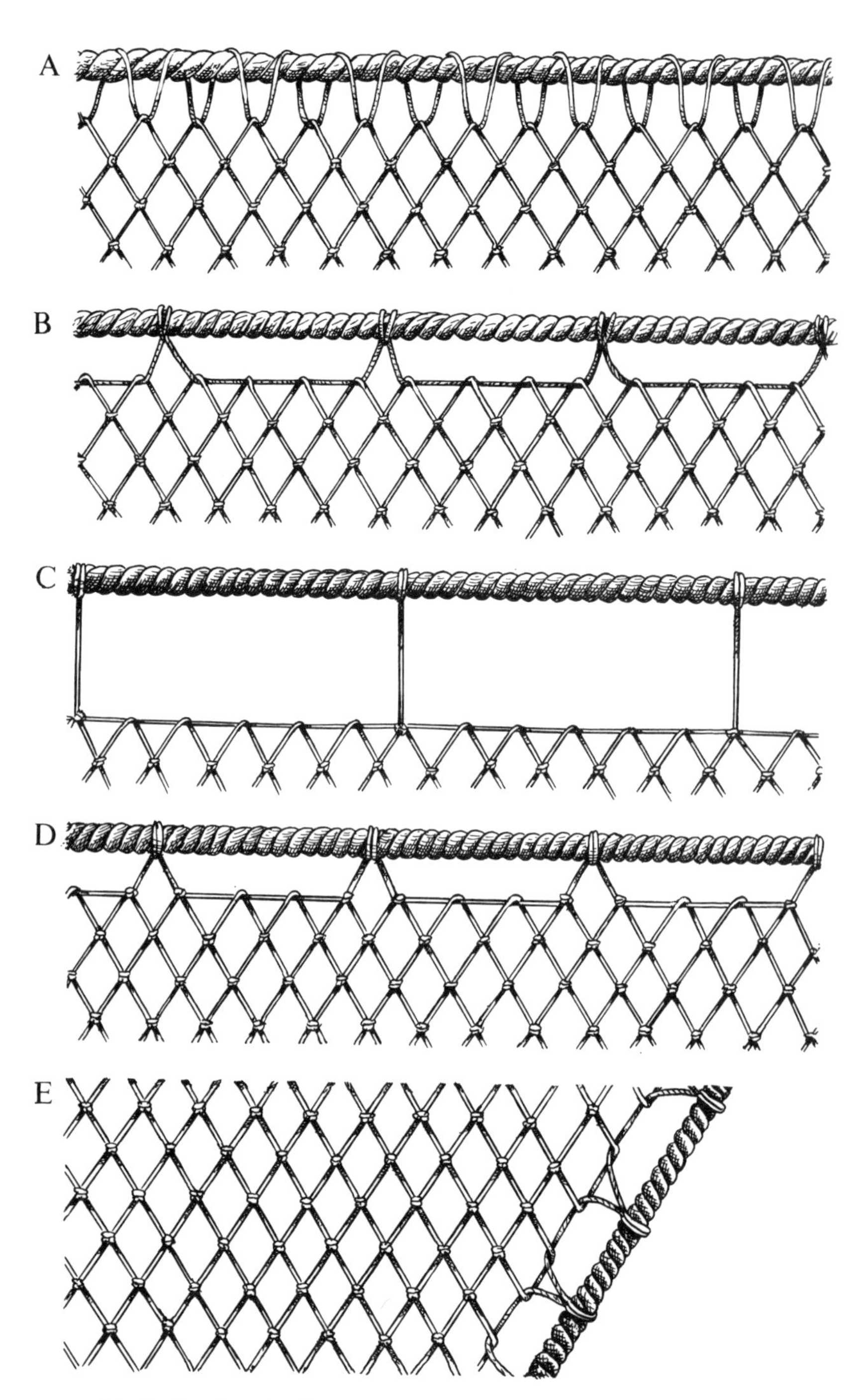

Methods of setting in a net: A reeving; B stapling by the third; C setting in by the third and norselling every sixth; D special method of setting in by the third; E stapling of fly meshes

ends to the head-line in the space where three would be fixed were the net stretched out tight. The meshes between the norsels are known as the "dormant meshes". In some nets the head meshes are supported by "running them through" with a light line before norselling . . . The fixing of selvage meshes is different. In some cases the end rope is simply rove through the selvage meshes, but usually the meshes are laced or marled to the rope, or stapled in a special manner, as in the fly-meshed wings of trawls. In the latter case one half of each adjacent fly-mesh is on adjacent staples, each fly-mesh having parts of adjacent staples passing through it.'

There are many variations, not only in the design of nets, but also in their setting, the variations being due not only to local traditions but also to the condition and configuration of various rivers. For example, due to the muddy nature of the banks of some rivers, such as the Wye and Usk, seine or draft nets cannot be used and salmon fishing has to be undertaken with stop nets, drift nets and basket traps. Although the shore seine is by far the most common instrument of commercial salmon fishing in English and Welsh rivers, many variations occur in the setting of those nets. In the Menai Straits in north Wales, for example, each seine net is set in by the fifth in its centre with the usual setting in by the third on its sides. This is to give the net a distinct bag-like characteristic in a stretch of water where currents are particularly strong. In the Teifi estuary, to quote another example, one seine netsman[16] describes the method of setting as follows: 'Mark the head-rope with the position of seven meshes, and set twelve meshes in the space marked by the extremities of those seven meshes. Take a stick and mark on it the position of those twelve meshes and transfer the marks to the remainder of the head-rope . . . Insert corks 14 inches apart on the head-rope and pieces of lead equidistantly on the foot-rope. Allow 2 feet or 3 feet more in the length of the foot-rope in relation to the head-rope to compensate for the unevenness of the river bottom. Bind the net to the foot-rope and do not tie a knot, as knots will make the net roll in use.'

Coracle nets can differ considerably from one river to the

other. Those used on the Teifi, for example, are by tradition made of hemp with the lines of horse-hair, while those on the Tywi are made of linen or hemp with lines of cow-hair. The former are equipped with pieces of lead attached at equal intervals to the foot-rope *(plwmffun)*, but in Tywi coracle nets lead weights are attached at irregular intervals to the foot-rope, their exact position being decided according to a most complex formula, dependent on the state of the tide and current at any one time. The Teifi nets are not corked, the Tywi nets are, and the Tywi netsmen seem far more conversant with the methods of adjusting the slackness of a net – closing or inserting a mesh according to the amount of water in the river – than those of the Teifi.

Nets made of non-synthetic materials have to be treated

Making a coracle net at Cenarth on River Teifi, circa 1938

before use for preservation and for colouring. In the past, tanning liquor obtained from a local tannery was the most common preservative. At Nefyn, Caernarvonshire, for example, the annual treatment of nets was carried out in a boiler on the sea shore, using tanning materials from a Caernarvon tannery. Until recently 'cutching' was carried out, for example, on the banks of the Wye at Chepstow and on the banks of the Teifi at St Dogmaels using a copper boiler embedded in a brick structure. By far the most common preservative is cutch, made from the wood of *Acacia catechu* of India and Burma ('true cutch') or from the bark of a mangrove ('Borneo cutch'). 'The process of cutching a net consists of either dipping it two or three times in the hot solution of "cutch water" and drying it each time, or of soaking it for considerable periods, say two days. After the first cutching of a new net, it is also dipped periodically in the solution, if cutch is the actual dressing for the net . . . The actual twine of a cutched net becomes thicker owing to that material becoming filled with cutch components that are insoluble in cold water.'[17]

A nineteenth-century textbook on fishing[18] describes the process of cutching as follows: 'For a net 24lb weight, previously untanned, take 4lb of pulverised catechu and boil it until thoroughly dissolved in 18 gallons of water, adding thereto, if procurable, about two hatfulls of young oak bark, pounded small and either put the net into the boiler, leaving it to steep two days and nights, or into a barrel and pour the hot liquid over it.'

In the past, both alum and coal tar were used for preserving nets, but since alum did not darken nets many fishermen believed it unsuitable. Others gave an opposite view, saying that white nets were far more efficient for catching fish, being more imperceptible in the water. To apply coal tar to a net, the method was to coil a net over a pole and dip it into hot tar for half a minute or so. 'Pass another pole through the coil, and two persons can wring it until all superfluous tar is pressed out, and those parts of the net which have escaped saturation in the dipping will receive it through the squeezing.'[19] Linseed oil, either on its own or combined with other substances such as red lead, ochre or paraffin, was also widely used for dressing nets in the past. Linseed oil was either boiled or applied raw and nets had to be soaked in it for four or six days. To avoid mildew, oiled nets were dipped in salt water. Of course, nets made of synthetic fibres need no pre-dressing.

For notes to Chapter 4, see page 309

5 Push Nets

Although small hand nets are less frequently used today than they were in the past, a surprising variety are still used in the various parts of Britain for the capture of salmon and sea trout. In the nineteen-thirties, push nets were used all the way round the coasts of the British Isles for the capture of shrimps, prawns and flounders, the principal ones being the Southport 'power net', the 'pandle' of the South Coast, the Lytham 'square net', the Cleethorpes 'shove net' and the Somerset 'skimming net'.[1]

The principal salmon push nets used on the rivers of England and Wales within the last quarter of a century are:

1. THE LAVE NET of the lower Severn, Usk and Wye. Variations of this net are also used in the Rivers Kent and Leven in north-east England.
2. THE LAMP NET of the Taf estuary in west Wales.
3. THE DIP NET of the River Parrett in Somerset.
4. THE CLICK NET of the River Humber, now disappeared due to the pollution of the estuary.
5. THE HAAF OR HEAVE NET of the Solway Firth and the River Lune in Lancashire.

The lave net is a Y-shaped structure, consisting of a hand-staff, known as a 'rock-staff'. This acts as a handle which the fisherman holds while the arms of the 'Y', called 'the rimes', act as a framework to the loosely hung net. The arms are hinged to the hand-staff, and are kept in position while fishing with a wooden spreader or 'yoke board'. Lave nets are, of course, used at low water and the estuaries of Monmouthshire were always regarded as their stronghold. Two licensees are still occasionally engaged in lave-net fishing at Dun Sands, near the mouth of the Wye, at Gruggy Rocks, but a total of six hold lave licences. During the 1969 fishing season 21 salmon, weighing a total of 235lb, were caught by lave nets at Gruggy Rocks. The nets were widely used for poaching on Dun Sands in the past. 'The men usually say they are after ray or skate, but you hardly catch anything on the Dun Sands except salmon', said one writer.[2] 'They're supposed to bring the salmon they catch to the slip and have it weighed, then they get a percentage. But they often throw salmon up on the bank . . . fetch them later on, and sell them to local people.' With the exception of the Severn, Usk and Wye estuaries, the only place where lave nets were used in Wales was in Carmarthen Bay, especially round Laugharne and St Clears, in the shallow estuary of the Taf. In those districts, the lave net was known as a 'lamp-net' but it never achieved popularity.[3] The lamp nets, which have not been used for at least the last ten years, resembled a Somerset dip net

Lave netting off Tites Point in the Severn estuary, 1965

rather than a true lave net, the net being attached to a wooden framework 3 feet wide and 3 feet 6 inches long, the cross piece at the back of the net being attached to a handle also 3 feet 6 inches long. The mouth of the net was the same width. The lamp net was used more in the manner of a lave net rather than a true heave net and could be used only in a very shallow estuary, such as that of the Taf.

There are limitations, enforced by the river authorities, on the size of lave nets. The Usk River Authority, for example, decrees[4] that 'The minimum size of the mesh . . . shall be 2¼ inches from knot to knot or 9 inches round the four sides measured when wet . . . Lave nets shall be of single netting, so constructed as to form a bag or purse suspended from a wooden frame, consisting of a pole or handstaff or handle, with two movable arms, each 5 feet 6 inches in length, and having a space of not more than 7 feet 6 inches between their outer extremities when fully extended, and the manner of using the same shall be by one person standing in the water and supporting or holding the net and lifting or scooping any fish that may be enclosed therein'. Regulations for lave net fishing on the Wye are exactly the same.[5]

J. N. Taylor[6] says that the origin of the net is obscure, 'but it is most probably the "ladenet" of John Smyth of Nibley, who, writing in 1639, says "Howbeit at certain places and seasons in lakes or pools (in the Severn) any stranger may with a Becknet or Ladenet, fish".' Closely related to the lave net was the so-called clenching net widely used on the Wye and Usk in the past for catching a variety of fish. A witness at the commissioners' meeting at Newport in September 1860, for example,[7] said of the Usk 'almost every farmer or farmer's lad has a clenching net, which is a net with a very small mesh put at the end of a long pole like a very large landing net; they slip it down the banks in the pools when the trout and other fish come to feed, and bring out sometimes a considerable quantity. Of course, they take anything, whether it is trout or salmon pink because the mesh is so small'. It is interesting to note that cleaching or clenching nets are still used on the Wye around

Symonds Yat for catching allis shadd *(alosa alosa)* and twaite shadd *(alosa fallax)* for a period of two weeks in May or June. Licences are issued for two weeks only and a water bailiff supervises the use of the nets.[8] Shrimp nets used in the Wye estuary at the present time are constructed on the same principle as the lave net. An example made by L. Till of Chepstow in 1970, for example,[9] has a hand-staff 42 inches long, a yoke board 16½ inches wide and two rimes, each 76½ inches long. The hand-staff is of ash and the rimes of hazel, and at the extremity of each rime a square piece of leather is nailed on the lower side. This is to prevent wear on the tips of the rimes and to facilitate sliding along the river bed, when the net is in use for shrimp catching. The mesh of the hand-braided net is no more than half an inch from knot to knot.

The use of the lave net for catching salmon demands considerable skill and agility, 'for with this simple gear the fisherman hunts the salmon in its own element'.[10] A fishing expedition may involve a fruitless wait of many hours with water up above the fisherman's knees. As soon as he sees the wake of a salmon as it rushes through the shallows, the fisherman runs to intercept its path. He lowers the net into the water just as the salmon approaches, and once it is over the headline and safely in the net, the fisherman has to step back smartly to counteract the force of a rapidly moving fish caught in the net. The hand-staff is raised and flipped over so as to enmesh the salmon in the fold of the net. If there is a convenient bank nearby, the net is grounded and the salmon killed with a wooden knocker or 'priest'. This tool, an essential for all salmon fishermen, is a wooden club, often of apple or holly wood, 8 inches to 12 inches long, usually attached to a lanyard round the fisherman's waist.

In the Oldbury district on the left-hand bank of the Severn, the method of lave netting is somewhat different. Due to the muddy nature of the river bed, it is impossible to wade in the water, and the fishermen utilise 'standings' for fishing. 'These standings are huge blocks of stone, set up centuries ago to command the outflowing stream of an ebb channel. At low tide the tops of these standings are exposed, and then the fishermen are to be seen standing on them, lave nets at the ready, watching keenly for the movement of salmon coming down stream.'[11] Since the standings at Salmon Pool, Oldbury, are some distance from the shore, the fishermen have to be very careful and watchful so that they leave the standings in good time before the flooding of the tide makes their journey back to the shore impossible. Flooding is sudden and fishermen always fish in pairs so that while one fishes, the other keeps a look-out for the incoming tide.

To carry salmon from the river, a number of devices are used. Sometimes a cord is threaded through the gills of the fish, and fish and cord are attached to the fisherman's waist belt for carrying. Sometimes, too, 'a salmon slip', a length of cord up to 7 feet in length, with a spliced loop at either end, is used to secure and carry the catch, the ends of the cord being looped over the salmon's tail and head. At Chepstow, a specially constructed carrying basket ('frail') is often used for carrying the fish. This is made of green willow, woven openly in the manner of a putcher and measures 20 inches or more in length. A slit is provided at the top of the basket for the insertion of the fish, and the basket is carried either with a couple of rope loops or loops of twisted withy. The baskets are made by the Chepstow putcher maker, L. Till.

Lave net fishing has virtually disappeared from the rivers that come under the jurisdiction of the Wye and Usk River Authorities, but in 1970 the Severn River Authority licensed 63 lave nets – 35 for the full season and 28 for a half season. A total of 1,076 salmon was caught by the fishermen using lave nets.

On the other side of the Bristol Channel, a certain amount of salmon fishing is carried out in the River Parrett by means of dip nets operated from boats. This is 'a simple bag of net supported between two withies on the end of a pole. (It is) used for dipping the salmon out of muddy water when on the surface'.[12] The Bye-Laws of the Somerset River Authority specify that 'the opening of the net shall be not more than

24 square feet in area with the mouth being not more than six feet from the heel of the net . . . The net shall fall freely from the opening without any further support, to the depth of not more than three feet below the line of the handle or pole'. These dip nets were designed for use in shallow water and in the nineteen-thirties they were used in the Severn estuary as well as in the tidal reaches of the Parrett. In 1968-9 the Somerset River Authority issued 12 dip net licences at £5 each but in 1970-1 the number of licences issued had declined to 3. In 1967, 25 salmon were caught by the netsmen; in 1968, 8 were caught and in 1969 only 4 salmon were caught. In the eighteen-sixties[13] every boat or barge that travelled up river between Bridgwater and Langport had a skim or dip net. 'The water is so thick, that the salmon swim on the top and when so doing, the fisherman pulls after the fish and, with a small net, used like a landing net, he skims or dips him out.' Today, dip netting is carried out with a net of cord mesh fixed to a stout pole. It is operated by a man standing in a small boat, known as a 'Parrett flatter', which is pointed at both ends, but it can be used for fishing from the river bank. 'The effectiveness of this net depends upon the fish swimming near the surface of the river as they come up on the flood tide. When the fishermen see the dorsal fin of the fish breaking the surface of the water, they guide their boat towards the fish, and if all goes well the fish is scooped into the net.'[14]

A net somewhat similar in design to the Somerset dip net was the so-called 'click net' used in the Humber. This was similar in shape and size to the Somerset dip net but it differed in that the click net had a complete metal frame, whereas the dip net is open-ended. The Fishery Bye-Laws of the Lincolnshire River Board[15] still specify that 'Stand, Bow, Click or Topping nets' may be used for taking salmon. Until about 1948 most of the salmon in the Humber were taken by these nets, 'taking the fish from the surface waters, where in the course of their migration, they are driven on account of the indigenous suspended solids in the rivers, known locally as "warp"'.[16] Click netting was carried out from a boat and was specifically designed for a river heavily polluted with waste. 'Fish migrating up the river were forced to the surface by the low dissolved oxygen in the water and the zone where oxygen levels are lowest coincides roughly with zones where warp is thickest; this gave a clue to the fishermen where the fish might be found. Where salmon reach the surface of the water in this way, known as "topping", they were in a more or less distressed condition and it was possible to row a boat up to them and dip them from the water with a large landing net, known as a click net. No click net licences have been sold in this area since the nineteen-fifties, but before the War it was quite a common method.'[17]

The 'heave' or 'haaf' net is a purse of netting fixed to a T-shaped wooden frame and is fished by one man. It has a limited distribution to north-west England and south-west Scotland. It is found in the River Nith in Dumfriesshire, on the Solway Firth and on the River Lune in Lancashire. Until recently it was in use on the River Wyre in Lancashire but, says the Fisheries Officer of the Lancashire River Authority,[18] 'More and more applicants for heave net licences appear each year, attracted no doubt by the good prices that salmon and sea trout are fetching, and the Conservators have now obtained a limitation order to prevent over-fishing. Some heave netsmen have moved up from the Wyre to the Lune estuary, the pollution in the former river having had an even worse effect than the pollution in the Lune.' The ten-year limitation order of 1953 was continued for a further ten years in 1963 by a further order of the Lancashire River Authority. At one time the net was also used on the River Ribble in Lancashire. In 1971 no fewer than 162 licences for haaf netting were issued by the Cumberland River Authority alone, for use in that part of the Solway Firth that comes under the jurisdiction of the Authority.

The haaf net is an unusual fishing instrument. 'It is virtually a kind of portable stake net of the bag-net type, being fixed to a wooden frame which is held by the fisherman on the bottom of the river.'[19] The net consists of a horizontal wooden frame of red cedar or pine known as a 'have beam' on the Lune and 'a bank' or 'baak' on the Nith. This varies in size from place

Haaf (heave) net of north-west England

to place and according to the individual fisherman's preferences. Small haaf nets with 8 foot beams, known as 'lifters', were in use in the past in the shallow waters of the Solway Firth, but those in use today vary from 14 feet to 18 feet in width. Running from the horizontal beam of the net and at right angles to it are three sticks, usually 4 feet in length on Solway nets and 3 feet in length on the Lune. In Scotland the sticks are called 'rungs', 'end rungs' and 'mid-rungs', but on the Lune they are called 'end sticks' and 'middle staff'. The end stick fits into sockets on the heave beam but the middle staff is a continuation of the handle. On the Scottish side of the Solway Firth the middle spar is usually off-set slightly to the right of the centre. The bolt rope of the net, which has a regulation mesh of 1¾ inches from knot to knot,[20] is lashed to the horizontal and the two end spars. It hangs freely across the rest of the frame.

The haaf net, says one report,[21] is 'a clumsier weapon' (than the lave net). 'The mouth is rectangular and measures about 16 feet by 4 feet. It has a handle in the middle of the long side.

The fisherman stands in the tide with the middle stick over his shoulder and the net streaming behind him. When he feels a fish strike, he lifts the lower tip to prevent its escape. This net is not easily manoeuvrable and there is not the same possibility of hunting as with the lave net. The fisherman must place himself within a few yards of the probable course of moving fish and await the event.' In describing the heave nets of the Lune, R. S. Johnson says,[22] 'Heave nets are fished on ebb and flood tides. The strands vary with the vagaries of the main channel. Some stands are good for the ebb only, others for the flood only, some for both. The netsman takes up his stand facing the stream, leaning on the handle of the heave beam, holding down on the frame, so that the net is ballooned out on each side of him by the current, and there he waits in silent meditation for a fish to strike the net. As soon as he feels a touch, he tilts the bottom of the frame up, and if he is quick enough, the fish is struggling in the net. Out comes the priest and the dead fish is hung by string through its gills on his belt. If he is fishing the flood, he holds his stance until the rising tide is almost up to his arm-pits, then shifts in-shore a yard or so and then starts again. On the ebb he moves out as the water falls. Sometimes a few of the heave netsmen will form a line from the shore out towards the centre of the channel, and in low water this forms a deadly barrier to the salmon or sea trout.'

Sanderson[23] describes the techniques of fishing with a haaf net as follows: 'The fisherman wades out into the river carrying his haaf-net on his shoulder. (On the Nith, where the net is deeper, the middle rung is offset from the centre to make the frame easier to balance.) When he reaches the right depth (often chest-high), the fisherman faces the tidal current, flow or ebb, and places his net in the water in front of him with its rungs resting on the bottom of the river. He stands with one foot forward to keep his balance in the current. With his right hand he grasps the central rung above the beam, and with the thumb of his left hand he gathers up six meshes of the net at the centre (or on the Lune a cord called the *finnel-cord* which gathers up the net), and then grasps the beam with his left hand just on the left of the middle spar. This makes the net stream out into two small bags or pokes on either side of him and slightly behind. On the Nith these pokes are about 8 feet 6 inches and on the Lune, where the net is only 3 feet deep, the pokes are about 6 feet.

'When a salmon runs into the net the fisherman takes a step backwards and either lifts his net out of the water or else pulls the mid-rung towards him, so that the frame floats quickly up to the surface. Meantime he flips the net over so as to enmesh the fish in a double fold of the net so that it cannot escape. He clubs the fish with a stick attached to a lanyard round his waist – called the *priest* on the Nith, the *nep* (nap) or the *killer* on the Lune – and has then caught and killed his fish. On the Nith the fish are put into a sack slung over the shoulder; on the Lune the fishermen carry a *fish-cord* or lanyard which is threaded through the gills of the fish and secured round the fisherman's waist so that the fish stream out in the water behind him.

'The fisherman then lifts his frame out of the water, shakes the net free, and takes up his stance again. With the small lifters there was one difference in technique: the fisherman could see the fish and pursue them across the shallow fords.

'Haaf-net fishermen sometimes operate singly, sometimes in teams standing side by side in order to cover a broad channel.'

Werner Kissling, in his description of Solway haaf nets,[24] says that 'The net, today, differs in no important respect from that used in the past. It is fixed to a beam or a cross-bar and to three "rungs" at right angles. There is one of these at each end and one, the midrung, not quite in the centre and projecting through the cross-bar (formerly known as the "haaf-bawk") by which the frame is held.

'Standing in the running tide as deep in the water as his breast-high waders will allow, the "haafer" sets his net against the stream and holds it firmly with his left hand on the centre of the beam, pressing it down and out and keeping it at arm's length, while reaching out with his right arm to grip the handle. The correct position for the fisher is to stand with one foot forward, so that he is able to lean against the fishing frame and yet, if necessary, quickly withdraw his weight and regain his balance.

'He then pulls up six meshes on his thumb so that a small bag or "poke" forms at each end of the net. When a fish strikes against one of the pokes, the haafer, feeling a pull at his thumb, instantly takes one step backwards, and presses down on the midrung. His haaf then floats to the top, ". . . from each corner of the net they have a warning string coming, which they hold in their hand, which gives them warning when the least fish comes in the net, and presently they pull the stakes (for so they term the frame of timber) from the ground, which are instantly wafted to the top of the water, and so catch the fish . . ." (*Stat. Acc Scot,* 1791: 15). If a salmon is caught in the right poke, he flings it with his right hand over into the double yarn, then turns with his back to the tide, knocks it over the head with his "mell" (a mallet) and throws it out of the double yarn. Then he lets the current swell the poke out, draws it in again to reach the fish, puts his finger into the gill, balances the fish over his shoulder, still holding onto the gill, and slips it head first into a bag slung over his shoulders.

'When more than three are fishing in the same place, the haafers stand in a row. They "pile" for positions before restarting in any place or fishing ground. The method is apparently derived from the ancient practice of "casting the mell" which is retained among the fishermen of Annan. As the tide rises and becomes too strong, the haafers move in from the deep end, one by one, and as the tide ebbs they move farther and farther down and continue till low water. After bagging the fish, the haafer resumes his position in the row without delay.'

'The halve net', says the Report of the Salmon Fisheries Inspectors in 1862,[25] 'is similar in form to a shrimp net, the men wade in with them in a line down the shallow part of the estuary and as the water deepens with the rising tide the outer man takes the inside place.'

The rivers in which heave nets are used all flow into the northern part of the Irish Sea, either into the Solway Firth or directly into the sea itself. 'Special conditions obtain in these waters: in each instance the land is flat towards the coast and the rivers reach the sea through channels which cut through sand or mud-flats, and which are affected by the regular flow and ebb of the tides. It is in these tidal channels that haaf-net fishing is practised, in waters which are for the most part muddy and where the fish cannot see the fishers at any great distance.'[26]

For notes to Chapter 5, see pages 309-13

6 Stop-Net Fishing

Stop-net fishing involves the use of a large, bag-like net suspended from two heavy poles that form a V-shaped frame. The stop-net is somewhat similar to a lave-net, and may indeed be a development from it, but it is much larger and is without a handstaff. Fishing for salmon is always done against the tide, and the heavy boats that are used for carrying the stop-nets have to be moored firmly, broadside to the flow of the tide. In some places, the boats take up station against heavy poles driven firmly into the bed of the river; alternatively, in wide streams the boats may be moored to steel cables, fixed permanently from the river bank. In narrower rivers, the stopping boats may be moored to ropes, that are fixed to stakes and anchors embedded in each bank, so that the boats and the nets that they carry can be moved at will, to any section of river.

Stop-nets are used today in three places only:

1. On the River Severn at Wellhouse Bay on the right bank of the river.
2. On the River Wye at Chepstow.
3. On the Eastern and Western Cleddau in Pembrokeshire.

In the nineteenth century, stop-nets were far more widely used than at present. On the Severn, for example, at least 16 boats were licensed to fish on both sides of the river while on the Wye, which may be regarded as the stronghold of this method of fishing, 37 stopping boats were licensed to fish in the Chepstow district alone. Stop-nets were in use as far up river as Lydbrook, not far from Ross (Grid Ref 604169). On the Usk, stop netting was widely practised in the mid-nineteenth century, but by 1902, according to the Usk Conservators' report, the number had declined to 7 while by 1913 stop netting had ceased on the Usk. On the Eastern and Western Cleddau, an observer in 1861 saw 'as many as 20 boats as close as they could moor without interfering with each other'.[1]

A stop-net consists of two spruce or larch poles, up to 25 feet long on the Wye and Severn, bolted or lashed together to form a V. The poles or 'rimes' give an entrance width of 30 feet, and the mesh of the net, which hangs loosely from the rimes, varies from 5½ inches on the headline, gradually decreasing in size to 2½ inches from knot to knot at the junction of the rimes. On the Cleddau, where the stop-nets are known as 'compass nets' because of their resemblance to a draughtsman's compasses, the rimes are shorter than on the Wye and Severn, and are usually no more than 19 feet 6 inches in length. The net entrance when the rimes are spread out is usually about 24 feet, and a 4 inch mesh throughout the net is used. In all cases the net is firmly lashed to the rimes with rope.

Across the rimes, about 6 feet from their junction, is fixed a light pole – 'the spreader', whose function is to keep the rimes apart when the net is in use. In some cases as on the Wye Boats, the spreader is adjustable and the exact position of the spreader on the rimes may be selected by pinning it through one of eight or more holes. Thus the width of net is adjustable. The spreader is fixed to the other rime through a metal collar, or is lashed to it with rope. When the net is not in use, the spreader is removed, and the net is wound around the rimes for storage. On the Wye, however, the net is set, with the spreader in place before the boat is sculled down-river or up-river from the boat-house at Chepstow. When fishing, a 'fly rope' attached to the boat prevents the net from going too deep and at too sharp an angle in the water. In Pembrokeshire the rimes are set in the fishing position after the boat has reached its allotted fishing berth.

The net attached to the rimes forms a loose bag and when fishing the ends of the rimes are pushed down into the stream, so that the whole net is under the surface of the water. On the Cleddau, the rimes will rest on the river bottom while fishing, but on the Severn and Wye, where the depth of water is so much greater, this is not necessarily the case. When the boat has been moored in its fishing position, the spreader is put in place and a forked stick four feet long is pushed under one of the rimes to keep the net in its correct fishing position. A loop of rope, known in Chepstow as a 'becket', is fitted to a ring in the centre of the boat. This is attached to the rimes when they are taken out of the water and not in the fishing position. The net is placed over the gunwale, so that it faces up-stream, against the ebbing tide. The fisherman takes up position against the downstream gunwale, with his back towards the net. Three or more 'bobbin strings' or 'babbling lines' are attached to the cod-end of the net, and when the net is in position, the fisherman gathers up these strings until he feels the net when the fish strikes. On the Wye, the bobbin strings, that usually number three, are attached to a piece of leather placed loosely in the fisherman's hand, but on the Severn, the strings which may number five or more are attached to an upright forked stick, known as a 'tuning fork'. This is similar in shape and size to the stick used to support the rimes of the net. On the Cleddau, the three bobbin strings from the cod-end of the net are merely held in the hand. In the swiftly flowing Wye, the stop-netsman always keeps another forked stick, called 'a kelp', close at hand, in order to push away any floating timber or debris that may enter the net and damage it.

With the net in its fishing position and the bobbin strings in his right hand, the fisherman waits until he feels a salmon strike the net. 'False alarms are fairly common. A submerged log or waterlogged bottle can enter the net undetected and cause the fisherman to knock out' (that is, to lift the net). 'Even a sudden shift in wind direction can trigger off a response . . . It is an old fisherman's law that "The man who knocks out most times with an empty net is also the man who will catch most fish." It indicates a greater sensitivity.'[2]

When a salmon hits a net, its impact is communicated along the cords to the fisherman's sensitive hand. He quickly kicks away the forked stick, that supports one of the rimes, and bears down on the end of the rimes, thus raising the mouth of the net out of the water. Heavy weights, known as a 'Man Friday' in the Chepstow district, that were placed at the junction of the rimes at the beginning of the fishing operation, act as a counter-balance in knocking out. The cod-end of the net is lifted on the down stream side of the boat. Should it contain a salmon, this is killed immediately with the wooden knocker that each fisher-man carries in his boat. A Severn and Wye stop-net is always equipped with a 'cunning hole', a 'trap-door' in the mesh of the net at its cod-end. This can be opened by undoing a cord so that the salmon can be extracted. On occasions the salmon may be removed from the top of the net, from the upstream side of the boat, rather than from the cod-end on the downstream side.

Stop-netting is a dangerous method of fishing for the in-experienced. To misjudge the strength of the tide may easily upset the boat, and throw the fisherman into his own net, when he would have little chance of surviving. It is said amongst Severn fishermen that one does not become a fully competent stop-netsman until at least one set of tackle has been lost in rough conditions on the river.

SEVERN STOP-NETS

On the Severn, as on the Wye, stop-nets are classified as fixed engines and were subject to the onus of proof of 'ancient right' required by the Special Commissioners on Salmon Fisheries in 1865. However, stop-net fishing has every appearance of being an early nineteenth-century invention.[3] The stopping boats themselves are broad-beamed, sturdy vessels, 35 feet in length, and are large and solid enough to withstand the considerable force of Severn tides. When fishing the boats are secured to steel hawsers, each one known locally as 'the chain'. The chains are secured to the river bank and extend for 200 yards to their anchorage in the sand of the river bed. The chain when in use passes around a pair of small capstans on the down-river gun-wale of each boat. In Wellhouse Bay, there are seven of these chains, with fishing rights for six boats on each of them. 'Today it is no longer profitable to operate more than two boats, however. Which chain they use will depend on prevailing wind and tide conditions, and the shape of the river bed.'[4] The six

Stop-netting in Wellhouse Bay, Severn estuary

fishing stations in use today have been recognised for well over a hundred years and the stations or berths were specified by the Commissioners for English Fisheries on 24 May 1866[5] as Hayward Rock, Long Ledge, Round Rock, Fish House, Old Dunns and Flood *(see pages 51-3)*.

In the eighteen-sixties, stop-net fishing was far more widespread in the Severn and certificates were issued for the following places:

Lydney Pill – 2 stop-nets (Owner – William Hiley Bathurst)

Coneybury Elm, Thornbury – 3 stop-nets (Owner – William Cullimore)

Etloe – 2 stop-nets (Owner – Edmund Barrow Evans)

Bull Rock – 3 stop-nets (Owner – Lord Fitzharding)

Wellhouse Bay – 6 stop-nets (Owner – William Hiley Bathurst)

The fishing season for stopping boats on the Severn extends from 2 February to 8 August with the river being closed from noon on Saturday and 6am on the following Monday. 'The fish run biggest in the opening weeks, but in recent years there has been a marked decline in the number of spring salmon, so that stop-netting is often not begun in earnest until the summer is well advanced. Returns are spasmodic. It is nothing to go several days without a fish, and then take a number from one tide . . . Ideal conditions for stop-netting are a 21 foot tide with a south-west wind. If the water is calm the fish can detect the presence of the boat and will turn away from it. As the tide ebbs, various currents are set up in the river. Salmon dislike a real strong current, preferring to swim in the slacker water alongside it.'[6]

WYE STOP-NETS

The Wye, according to the Commissioners on Salmon Fisheries in 1861,[7] is 'a large and noble river (and) possesses great capabilities for salmon fishing. Here, as on the Severn, the tide rises very high and extends for a long distance upwards, but the tidal portion scarcely anywhere assumes the character of an estuary; it is narrow and is filled when the tide rises with a deep muddy stream, which leaves steep sludgy banks at low water, unfit or at least inconvenient for fishing with a draft or seine net. The muddy water here renders nets of different kinds, held by hand or temporarily fixed during a part of the ebb or flow of the tide efficient, and it is by means of coracles or stop-nets that the salmon fishing in the lower parts is almost entirely carried on.' Undoubtedly the Wye has always been the stronghold of stop-net fishing, for in the eighteen-sixties no fewer than 37 stopping boats were licensed to fish in the Wye, in the vicinity of Chepstow alone. The swiftly flowing, mud-banked Wye, with its appreciable tidal range in the narrow gorge-like section of the river, makes it difficult to use any other fishing instrument but the stop-net. Lower down, near the confluence of the Wye with the Severn, both putcher weirs and drift nets are used, while up-river from Chepstow in the vicinity of Tintern and Chepstow both shore seines and weirs or cribs were used for salmon fishing in the past, in addition to the stopping boats.

The Inspectors of Salmon Fisheries in their *First Annual Report* of 1861 mention the Wye as an important river for stop-net fishing. 'Stone piers have been erected, jutting into the river of like form and purpose to the cairns . . . and stop-nets take the fish led into them by the eddy so formed. At Lydbrook, boards are placed upon an old weir stretched nearly across the river, and in certain states of water, every fish is captured by the stop-net below.' The Inspectors estimated that there were 23 stop-nets between the sea and Chepstow bridge, 15 between Chepstow and Llandogo, and 40 between Monmouth and Hereford. It was impossible to estimate the number of stop-nets between Llandogo and Monmouth, but the Inspectors expressed the opinion that the Wye was 'principally fished by stop-nets'.

From the dissolution of the monasteries to the present century, fishing rights on the Wye were in the hands of the Dukes of Beaufort, who leased the fishing to other people. The most notable of nineteenth-century Wye lessees was a Scotsman, Alexander Miller (1831-92), who built the Chepstow Fish House in 1864.[8] In 1901 the Duke of Beaufort sold the fishing rights to the Crown who leased them to the first members that formed the Wye Fishery Association. In 1928 the sole rights were transferred to the Wye Fishery Board, and that Board, together with its successor the Fishery Department of the Wye River Authority, have been responsible for all the stop netting, drift netting and putcher fishing in the river.

The seven stopping boats owned by the Wye River Authority today operate in definite stations or 'berths'. Below the railway bridge are 'Gut', 'Gloucester Hole' and 'Slip and Cliff'. Between the railway bridge and road bridge at Chepstow are 'Port Walls' and 'Behind the Plane', while above the road bridge are 'Inside the Sand Beds', 'Inside Castle', 'Outside Castle' and 'Bottom of the Hope'. Lots are drawn to decide which boat should go to which berth and a strict rota is followed after the initial drawing of lots. In practice, however, it is rare that more than two or three boats are found on the river at any one time, for all but one of the stop net fishermen are part-time workers, and the River Authority find ever-increasing difficulty in obtaining labour. The fishermen are paid a piece-rate wage. No longer does stop-net fishing attract a large number of people, who, in the past, regularly gave up their employment on 1 February each year to join the fishing teams.

In the nineteenth century the Wye stop-net fishing grounds were extensive and in addition to boats licensed to fish in the vicinity of Chepstow, designated 'The Lower Fishery' by the Commissioners for English Fisheries in 1863, stopping boats were also found as far up river as Monmouth and even as far up river as Lydbrook. Fishing rights throughout the river were owned by the Duke of Beaufort, and the licence for the Lower Fishery drawn up in 1866 says: 'We . . . certify that the said Fixed Engines consist of thirty-seven Boats with stop-nets used in the following places to wit – Twenty-three boats with stop-nets are used in sixty-two several Berths or stations in the Fishery known as the Lower Fishery between Chepstow Bridge and the Black Rock in the Severn on the West side of the mouth of the River Wye –

'Fourteen boats and stop-nets are used between the said Mouth of the River Wye and Cone Pill in the Severn, which is on the east side of the mouth of the River Wye, in the following places to wit – Four Boats and Stop Nets at or near Pill House – Nine Boats and Stop Nets in Beechley Bay and Five Boats and Stop Nets at and within one mile of Horse Pill – Four of the said nine boats used in Beechley Bay being also used within a mile further up the Severn and Four of the said Fourteen Boats and Stop Nets being used also at Chapel House on the left bank of the Severn.'

Above Chepstow road bridge, the Wye Upper Fishery licensed in 1866 possessed thirteen boats and stop-nets. All these were associated with cribs and artificial weirs built from the bank of the stream. The berths were Turk's Hole, Coed Ithol, Fleming's Weir (2 stopping boats), Brock Weir, Folly Crib (near Tintern), Crib (near Tintern Abbey), Great Ead, Plumb Weir, Inkins and Wall Weir (near Flouncer).

Additional berths were available at Passage Slip, Lancant Slip, Windcliff, Trooper's Hole, Chit Weir, Upper Rocks (2 berths), Slack of the Hope, Butcher Pitch (2 berths), Cinder Pitch (2 berths), Mudny's Hole, Rock above Tintern Abbey, Elms (2 berths), Brook (2 berths), Pride Lane and Shoal (2 berths), Brow and Pippin (2 berths), Fleming's and Mathis (2 berths),

Orls (2 berths), Old Barge, Hammer and Tongs, Elms, Peter's Thumb, Punreen, Fish House, Barclay's Slip, Cottrells, Slack of the Hope timber shoot (2 berths), Bottom of the Hope (2 berths), Little Sand Bed, Yew Tree, Sand Beds (2 berths), Castle (3 berths) and Castle Rock (3 berths).

Stop-nets and cribs were also used further up river at the Redbrook Fishery with two boats in four berths at Joe Surley, Hawthorn Tree, Biggs Weir Stones and Slackwater. At the Monmouth Fishery between Monmouth Bridge and Upper Redbrook three boats with stop-nets and cribs were in use. The berths were designated Fishinam, Little Crib, and Halfway House or Sentry.

The Wye stopping boats, although broad-beamed and sturdily constructed, are slightly shorter than the Severn variety, each measuring 25 feet in length, and with a 7 foot beam. The boats are kept in the River Authority's boat-house on the river bank at Chepstow during the closed season, but are taken down to the river and anchored in the stream below the boat-house for the duration of the fishing season that extends from 2 February to 15 August.

Each boat is equipped with a single oar 16 feet long with a blade from 5 to 7 inches wide for sculling and 3 iron-pointed stakes, from 22 feet to 30 feet long, which are used for mooring the boat in its fishing position in the river. When the boat arrives at its allotted berth, an anchor is thrown into the water over the bow of the boat, and the heavy vessel is rowed into such a position that it lies broadside to the flow of the tide. The manoeuvre demands considerable skill and strength, for the ebbing tide is very strong, and mooring a boat broadside against it is a most difficult operation. Two poles, one at the stern and a slightly longer one at the bow of the boat, are driven into the water on the down-river side of the boat. These are inserted at a sharp angle of about 75 degrees. Each pole may have to be driven into the river bed, half a dozen times or more, before the fisherman is satisfied that they hold the boat against the force of the tide. After the poles have been lashed to the stern and bow lashings on the gunwale, a third pole, designed to prevent any further movement, is driven into the river bed over the stern of the boat at an approximate angle of 45 degrees. This in turn is lashed to the gunwale and to the other stern pole. The operation of mooring a stopping boat on the Wye may take a considerable time, before the fisherman is perfectly satisfied that his boat is moored in the correct position and the stop-net can be inserted in the water.

On the Wye, stop-netting is an efficient method of fishing, particularly in a dry season, for 'low water netting conditions are a pre-requisite for a good netting season'.[9] In 1969 for example, when only four stopping boats were in use at Chepstow, a total of 276 salmon was caught. This represented 35.2 per cent of the salmon caught by the various types of instrument owned by the Authority; the drift nets and the Beachley putcher weir being responsible for the remainder of the catch of 783 fish.[10]

Until 1935 stop-nets were in use on the Usk, but in that year the only stop-net on the river at Caerleon belonging to the River Usk Stop Net Trustees was not let.[11] The trustees, says the Report, 'are now considering the advisableness of not letting in future any of the nets (there are 16, of which the Trustees are owners of 14 and 2 belong to Lord Tredegar, unless any attempt is made – which is very unlikely – on the part of Lord Tredegar to let the two nets which belong to him). The catches or pitches for these nets, which were formerly rented by the Stop Net Trustees, are now of no commercial value. It would be satisfactory if we could feel that all stop net fishing on the River had been done away with!'

Earlier in the century appreciable quantities of salmon were caught by stop netsmen, principally between Newport and Caerleon. Between 1910 and 1935 the number landed was as follows:

1910 - 603	1916 - 122	1922 - 89	1928 - 44	1934 - 3
1911 - 116	1917 - 225	1923 - 22	1929 - 10	1935 - 0
1912 - 403	1918 - 64	1924 - 160	1930 - 81	
1913 - 198	1919 - 16	1925 - 40	1931 - 49	
1914 - 163	1920 - 110	1926 - 96	1932 - 2	
1915 - 184	1921 - 32	1927 - 41	1933 - 2	

CLEDDAU COMPASS NETS

On the Eastern and Western Cleddau that run into Milford Haven in Pembrokeshire, the stop-nets that are used for catching both salmon and sewin are termed 'compass nets'. Stop-net fishing is said to have been introduced into Pembrokeshire by two Gloucestershire men who came from the Forest of Dean in the early nineteenth century to work at the Land-shipping anthracite mine. Two fishing berths on the Eastern Cleddau still bear the names Ormond and Edwards – the two Gloucestershire men who are said to have brought stop-net fishing to Pembrokeshire.

Throughout the nineteenth century compass netting was undoubtedly widely practised on the Cleddau rivers, and according to the evidence given to the Commissioners on Salmon Fisheries in 1861,[12] 'As far as salmon and sewin are concerned, the only fishing that takes any considerable number is by what are called compass nets. They fish with them by mooring boats across the river; they are about 20 to 25 feet in spread.' Unlike the Wye stopping boats that are moored against poles driven into the river bed, and unlike the Severn stopping boats, that are moored to permanently fixed chains, the Cleddau compass boats are not classified as fixed engines. The boats may, with ease, move their position along a rope or 'warp' fixed on either side of the river. In the past, the classification of Cleddau compass nets and movable fishing equipment led to considerable controversy, for the authorities maintained that they were fixed engines, as were the Wye and Severn stopping boats. In 1869 and in 1939, for example, compass nets were completely banned from the river. The 21 licensees who were allowed to fish at the time fought the issue, and compass netting was allowed once again. Nevertheless there was a rapid decline in the number of licensees, so that today only eight compass boats are allowed on the rivers. All the fishermen are residents of the small river-side villages of Hook and Llangwm.[13] The Hook fishermen, by tradition, operate in the Western Cleddau only, but the Llangwm fishermen are allowed to take their boats to both the Western and Eastern Cleddau.

The boats used for compass netting are ordinary 14 foot,

tarred boats that are used both for herring drifting and seine netting as well as compass netting. Each boat is approximately 5 feet in beam. Most of them were built locally by the fishermen themselves during the last quarter of the nineteenth century, and although the boats are small, they are sturdy enough to withstand the force of the ebb tide on the Cleddau.[14] Of course, the force of the tide is nowhere as strong as it is on the Wye and Severn. The fishing season extends from 1 March to 31 August, although few compass netsmen venture on the water before mid-May. The boats are moored at Little Milford Quay, 'the long poles of the compass nets protruding over the transoms like giant bowsprits in reverse'.[15] Orange-coloured synthetic material is used for making Cleddau compass nets and a four-inch mesh throughout is used. The net is 120 meshes wide at the head and 70 meshes wide at the tail and is between 6 feet and 7 feet in depth. It is attached to two larch poles, each 19 feet 6 inches in length, the sides of the net being approximately 100 inches deep, but forming a loose bag in the centre between the poles. It was customary to bury larch poles in the riverside mud for some years before they were considered ready for use.

For fishing, the boats are attached to iron or wooden stakes embedded in the river bank, and a fishing session can begin as soon as the stakes appear above the surface of the water, approximately two hours after high tide. A rough shelter equipped with a wooden bench is found on the river bank at Little Milford, and the fishermen wait in this until the stakes appear above the water.

A strict rota is followed to determine fishing stations. The first team to arrive at the fishing station at Little Milford begins at stake No 1 – The Bite, while at the next tide it has to go to stake No 2 – The Lake, proceeding to the other stakes on the following tides until all the berths are fished. The stakes on the Western Cleddau are The Bite, First, Second and Third at Lake, The Quay, The Drot, Level Stake, The Stones, The Rail (where the stake is a piece of old rail) and The Grim Bank. On the Eastern Cleddau, the stakes are Home Stake, Kerlisky, Layer's Park, Ormond, Edwards and Crafty.

Compass netsmen waiting for the tide on the River Cleddau at Little Milford, Pembrokeshire

Nicholls describes the method of fishing with one of the Llangwm fishermen as follows: '. . . the sturdy Reg Jones began his methodical preparations. Taking one end of his warp he made a bowline and dropped it over a short black stake, planted a little way down the steep sculptured mud of the bank. "All these stakes you see, comes up two hours after high water." . . . Then he was in his boat striking out sharply upstream in the slack water, while the warp, neatly coiled in the stern sheets paid itself out. When he reached midstream I could see how fierce the tide was; all the contents of those wide stately reaches of the river above us had to be sucked out through this narrow gap in the next three hours, and now the ebb was at its most vicious. It was only by a stern effort that he gained the other shore at a point opposite. Once there he produced the other end of his warp; attached to it was an anchor, one fluke of which he dug firmly into the mud a little way above the water.'[16] The boat is secured along the warp or rope which measures about 44 fathoms through a hook low down inside the boat, and held at the bow by a raised stem piece. A thole pin holds the warp in place at the stern. In operation, the boat and its net can be taken backwards and forwards along the warp, if the wave of a salmon is observed.

When the chosen fishing position along the warp has been decided upon, the compass net is inserted in the water, after the depth of water has been tested with the oar. That depth must not be more than the depth of the leather tacked around the oar in the position that it should be in the row-lock. In practice this is approximately five feet in depth. This is quite a spectacular operation especially on the Eastern Cleddau, when the force of the ebbing tide is said to be much greater than on the Western. The boat is rocked violently from side to side with the fisherman grasping the net poles, which are counterbalanced with one or two flat stones or concrete blocks at the apex. A final lunge, until the gunwale of the boat is within an inch or two of the

A compass netsman inserting the net in the water of the River Cleddau, Pembrokeshire, 1972

water, causes the points of the poles to slide down to the river bed. Three feeler lines run from the cod-end of the net and are held in the fisherman's left hand as he sits on the seat in the centre of the boat below the apex of the compass net. His right hand rests on the poles. As soon as a fish strikes the poles and net are lifted and the fish removed. Cleddau compass nets are not equipped with openings in the cod-end as those of the Wye and Severn, and the salmon are removed through the mouth of the net on the up-river side of the boat. Occasionally some of the fishermen fish after the tide has turned and set their nets facing the incoming flow tide.

The eight licensees of Hook and Llangwm each appoints an endorsee who can use the licence in the fisherman's absence. This ensures the continuity of compass netting on the Eastern and Western Cleddau. There has, of course, been a sharp decline in the number of fishermen, who are only able to fish between midday on Monday and 6am on Saturday. When more compass nets were in use on the river, says one fisherman,[17] 'I can remember when th'old folks used to what they called "stem" – come up here three hours before high water for to get the best stakes'. In this way many of them waited for six hours before they could begin fishing.

For notes to Chapter 6, see page 313

7 Coracle Fishing

The use of the coracle as a fishing craft on Welsh rivers has declined very rapidly in recent years, so that today coracle fishing is limited to three west Wales rivers – the Teifi, Tywi and Taf. Even on those rivers there has been a sharp decrease in coracle fishing. For example, on the Teifi, the picturesque village of Cenarth has long been regarded as the home of coracle fishing, but in 1972 no coracle fishermen at all operated from that village, the last licensee being unable to work after 1970. In 1807 there were so many coraclemen at Cenarth that a contemporary observer wrote:[1] 'There [is] scarcely a cottage in the neighbourhood without its coracle hanging by the door.'

The coracle, which is a keel-less, bowl-shaped fishing boat, has been known for many centuries in Wales and it may be regarded as the direct successor to the small skin-covered boats described in detail by Caesar,[2] Pliny[3] and other Roman writers.[4] It seems, however, that the vessels described by Roman writers as well as those mentioned in the *Mabinogion*[5] were sea-going, keeled boats, similar to the curraghs of Ireland,[6] rather than to the keel-less fishing coracles of Welsh rivers, that are specifically designed for operation in swiftly-flowing streams.

It is clear that in addition to the curragh type of skin-covered boat in use in the Dark and Middle Ages, fishing coracles were also in use. Little is known about their design, but in the *Gododdin* poem of Aneirin, which can be dated to the seventh century, one line reads *ef lledi bysc yng corwc*[7] (he would kill a fish in his coracle), while the medieval Welsh laws gives the value of a coracle as eight pence *(corwc wyth keinhawc kyfreith).*[8] The first clear description of the true river coracle appears in the writings of Giraldus Cambrensis, who in 1188 made a journey through Wales. 'To fish or cross streams', he says,[9] 'they use boats made of willow, not oblong nor pointed at either end, but almost circular or rather in the form of a triangle, covered without but not within[10] with raw hides. When a salmon thrown into one of these boats strikes it fiercely with its tail, it often oversets it and endangers both vessel and boatman. In a clumsy manner, in going to or coming from a river, the fishermen carry these boats on their shoulders.' Unfortunately, Giraldus gives no key as to where in Wales he encountered coracles were very common indeed at Cenarth and a witness to considerable detail in describing the salmon leap at Cenarth and the salmon fishing weir at Cilgerran and indeed he even describes the beavers that lived near the river at the time but he makes no mention of coracles at all. George Owen,[11] in the early seventeenth century, gives many details of fishing weirs at Cilgerran and seine nets in the estuary but nowhere does he mention coracle fishing, though he describes the river at Cenarth in considerable detail. By the end of the eighteenth century coracles were very common indeed at Cenarth and a witness to the Royal Commission in 1863 strengthens the view that coracles were unknown on the Teifi until the latter part of the eighteenth century.[12] The witness said that coracles 'have increased since I have known Cenarth, which is the principal coracle station . . . Coracle fishing has not been introduced on the Teifi from what I can gather above 60 years or something of that sort'. The absence of evidence relating to the use of coracles for fishing in the rivers of south-west Wales before the second half of the eighteenth century is surprising. But in north Wales, especially on the Dee, coracle fishing was undoubtedly commonplace in earlier centuries. Medieval *cywyddau* give some indication of the design of coracles in north-east Wales. In Cardiff MSS.64 and 12[13] they are mentioned. The first poem is by Ifan Fychan ab Ifan ab Adda soliciting a coracle from Sion Eutun. The second is a reply by Maredudd ap Rhys, who flourished between 1430 and 1450, on behalf of Sion Eutun. The third *cywydd* contains little relevant material on the design of coracles in north east Wales.

Cywydd 1

Am gwrwgl i ymguriaw
For a coracle to beat about

Am y pysg drud cyn y Pasg draw
For the valuable fish before next Easter

Crair lleder, croyw air Lladin
A leather relic, pure Latin word

Codrwyn du, caeadrwym din . . .
Black covering enclosing its bottom

Cod groenddu da, ceidw grinddellt
A bag of black skin, preserves dry laths

Y gerwyn deg o groen du
A fair vat of black skin

Bwcled sad, ble cela' son
A firm buckler (why should I conceal it)

Bas ydyw o bais eidion
It is shallow, made of a bullock's tunic

Padell ar ddwr ni'm pydra
A pan which will not cause me rot in the water

O groen cu eidion du da
Made from the fair skin of a good black bullock

Cywydd 2

Bola croen ar waith bual crwn
A skin bag in the shape of a circular horn

Blwch byrflew tondew tindwn
A short-haired, skinned box with broken bottom
(ie laths)

Nofiwr o groen anifail
A swimmer of animal skin

Noe serchog foliog o fail . . .
A fond, paunchy vessel

Llestr rhwth fal crwth fola croen
An open vessel like a skin bellied *crwth*

Coflaid o ledryn cyflo . . .
An armful of leather in calf

Myn Pedr, mae yn y lledryn
By Peter, there is in the leather

Rywiogaeth wyll a dwyll dyn.
The nature of a fiend that deceives man

A elai'r cwrwgl dulwyd
Would the dark grey-black coracle

I'r llyn a'r pysgotwr llwyd?
Take the fisherman to the pool?

Er dim ni ddeuai o'r dwr
Never would it come home from the water

Heb ysgwd i'w bysgodwr . . .
Until it had given a push to its fisherman

O'ch Fair, pan na chai efo
By Mary! Why does not he

Long o groen newydd flingo?
Have a ship of newly flayed skin

Groen buwch ar waith gweren bert
The skin of a cow worked with the fair tallow candle

One of the great advantages of the coracle over other types of fishing craft is its manoeuvrability and the fact that it only draws three or four inches of water. In shallow or rock-strewn rivers it is particularly useful not only for netting but also for angling. In eighteenth-century Monmouthshire, for example, the fishermen made use of 'a thing called Thorrocle or Truckle'[14] when fly-fishing for grayling. In the Dee, the coracle was regarded as essential 'on the rough, rocky middle reaches . . . when, owing to the force of the current and deep hidden ledges and clefts in the rocky bottom, wading is impossible in many places, and no other type of craft, not even a birch bark could possibly be used . . . A single coracle weighs some 30lb and a double one some 10lb heavier. In one of these it is possible to shoot rapids and dodge in between out-jutting ledges in the fastest and wildest stream, holding on by gaff or paddle to some outcropping ledge or rock, and one can turn oneself in perfect safety with a rush of wild white water on each side. By using a coracle one can therefore fish places that could never be reached either by wading or throwing the longest line from the bank'.[15]

Of course the coracle has always been used primarily for netting and both its manoeuvrability and lightness were important considerations in its persistence. The coracle fisherman usually has to carry the coracle on his back for considerable distances, for fishing is undertaken by drifting with the flow of river. 'It happened frequently', said a Report of 1861,[16] 'that several hundreds of men would go out very early in the morning with coracles on their backs, pass over the mountain and come some distance down the river, taking all they could catch with very fine nets.' Describing the lightness of the Monmouthshire coracles, Hawkins[17] says the coracle was so light that 'the countrymen will hang it on their heads like a hood, and so travel with a small paddle which serves for a stick till they come to a river, and then they launch it and step in; there is great difficulty in getting into one of those Truckles, for the instant you touch it with your foot it flies from you and when you are in, the least inclination of the body oversets it'.

When animal hides were used for covering the wooden framework of a coracle, the vessel was undoubtedly much heavier than more recent ones where flannel or canvas was used as a covering. 'Presumably it was when coracles were hide covered that the old Welsh adage took form which runs "A man's load is his coracle" *(Llwyth gŵr ei gwrwgl).* Today a coracle in south Wales seldom weighs as much as 30lb, surely a trivial load to carry. A hide-covered coracle would weigh nearly double this and would justify the proverb more fittingly.'[18] Until at least the early seventeenth century, the latticed wooden framework of a coracle was covered with horse or ox hide, the covering area of one hide governing the size of a one-man coracle. In some parts of Wales, particularly south-west Wales, flannel had replaced animal hides as a covering by the end of the eighteenth century and continued to be used until the mid-nineteenth century.

It has been suggested[19] that canvas was a cheap substitute for flannel, but even so according to Donovan 'flannel was of a more durable substance, [it] may be easier prepared and keeps out the water much longer than canvas'.[20] In the Llanegwad Vestry Book on 7 September 1798, the following entry appears: 'that John Harry, overseer, do purchase flannel and other things necessary to make a coracle for John Lot.'[21] During the first decade of the nineteenth century, however, skin-covered coracles were still to be seen, but they were rare[22] for 'a kind of coarse Welsh flannel . . . is generally made use of. The particular sort of flannel proper for the purpose, could be purchased a few years past at a low price, but it is at present worth two shillings a yard upon the spot where manufactured, and hence through notions of economy, canvass prepared in the same manner is becoming rather more universal than before'.

J. R. Phillips in 1867[23] says that flannel was used 'until recently' for covering coracles at Cilgerran, while a Cenarth coracle fisherman interviewed in 1961[24] said that his father used flannel for this purpose until about 1880.

This was obtained from the woollen mills of Dre-fach, Carmarthenshire,[25] and it was prepared by dipping in a boiler containing a mixture of tar and rosin.

When a flannel was fully saturated, it was taken out of the boiler by four men, one at each corner, and laid down on the upturned coracle frame and tacked into place. This seems to have been the usual method of covering adopted throughout south-west Wales, but it is unlikely that flannel was used in other districts. Pennant, in his tour of North Wales,[26] remarks that coracles on the Dee 'have now lost the cause of their name, being no longer covered with *coria* or hides, but with strong pitched canvas', while Bingley, ten years before, remarks that Dyfi coracles too were covered with canvas rather than hide.[27]

On a print at the Carmarthen Museum, dating from 1794, some indication of the design and covering of a Tywi coracle appears in a verse that reads:

Upon the glittering stream below,
Those fishermen of courage bold,
In numerous pairs, pursue their trade
In coracles themselves have made;
Form'd of slight twigs with flannel cas'd
O'er which three coats of tar are plac'd
And (as a porter bears his pack)
Each mounts his vessel on his back.

Nevertheless, for at least the last eighty years, Welsh coracles have been canvas or calico covered, and flannel-coated vessels with three coats of tar have ceased to be used. The method of covering on the Teifi and Tywi is to use unbleached, twill calico, 5 yards long and a yard wide, cut and then sewn up the middle. This is tacked to the coracle frame and 6lb of pitch mixed with half a pint of linseed oil is then boiled thoroughly, allowed to cool and then applied to the outside of the coracle. Preferably this is done in the open air on a warm day and a single coat of pitch is usually sufficient to make the coracle waterproof. In Cardiganshire and Carmarthenshire, covering coracles is usually regarded as a woman s duty.

The life of coracles is indefinite, although many Teifi fishermen believed they should obtain a replacement every two years. Slight damage to the craft was repaired by applying a tarred patch over a tear or hole, but if repair was not possible, it was quite common for a fisherman to re-cover an old coracle frame with a new piece of pitched canvas or calico. To repair a hole or tear in the covering of a coracle a hot poker was applied to the pitch surrounding the tear until it had melted. A tarred calico patch was then applied over the tear and the hot poker used again to spread the pitch over the patch. If by any chance the coracle was holed while on the water, a piece of lard, which was always carried in the coracle, was stuck over the tear as a temporary repair.

The design of coracles and the methods of using them vary considerably from river to river. They differ according to the physical nature of the individual streams – whether a river be swiftly flowing or slow moving, whether it has rapids and much rough water and whether it is shallow or deep. Design varies too according to the preferences of the individual fisherman and whether a fisherman prefers a heavy or light coracle. A Teifi coracle, for example, can weigh as little as 25lb and as much as 36lb, while length varies from 50 inches to 60 inches, the actual size and weight depending on the preference of the fisherman. Design varies too, according to the tradition of the various rivers, for in Wales a remarkable homogeneity in the design of coracles occurred on the various rivers, and distinct regional types were in existence for many centuries. For example, although the coracles of the Tywi and the nearby Taf are somewhat similar in shape, the Taf coracle, designed for use in a fairly narrow, swiftly flowing stream, is heavier than the Tywi variety. Instead of the wattled gunwale of the latter, it has a planked gunwale. The Taf coracle is sharper at the fore end and flatter at the

A coracle builder at Cenarth on the River Teifi, circa 1938

other, and usually weighs about 33lb compared with a maximum weight of 28lb in Tywi coracles. During a recent fishing season, when a Taf licensee was unable to make a coracle for his own use, a Tywi coracle from Carmarthen was borrowed for the season. The Taf fisherman was not happy with the design and performance of this vessel on the river and he soon reverted to the traditional coracle built by the craftsmen of Lower St Clears, specifically for the Taf.

The Salmon and Freshwater Fisheries Act of 1923 put an end to coracle fishing on many rivers such as the Severn, and severely restricted the use of coracles on many others. Subsequent river authority bye-laws have caused an even more rapid decline in coracle fishing and in some districts the coracle is on the point of disappearing. For example at Cenarth on the River Teifi, long regarded as a centre of coracle fishing, legislation in 1935 prohibited the issue of new licences to fishermen in the non-tidal section of the river above Llechryd bridge. Consequently, the number of coracle licensees has declined so that today no coracle fishing at all is practised at Cenarth. It was estimated in 1861 that there were over 300 coracles on the river with about 28 pairs of fishermen in full-time occupation as coracle fishermen above Llechryd Bridge.[28]

Below Llechryd Bridge however, where the river is classified as tidal, there is no restriction and five pairs of coracles are worked on a part-time basis in the picturesque Cilgerran gorge. The coracle men there use both the traditional, armoured coracle net and the now illegal set nets for catching salmon. The gorge is an ideal place for using illegal equipment,[29] for it cannot be reached by road or footpath. A newspaper report, for example, notes that for the period June-October 1969, '81 illegal set nets laid by "latter-day pirates" were removed from Cilgerran gorge'.[30] The set net *(rhwyd fach)* is a single, unarmoured net 18 feet to 50 feet long, usually of fine mesh[31] attached to the river bank by means of a light stone, and the other end is spread out as far as it reaches into the river. Lead is fastened to the foot-rope and corks to the headrope. The net is usually placed in the water by a coracle man and it can only be used in fairly still water. 'Although corks are attached to the top of the net', says the 1971 *Report of the South West Wales River Authority*, 'there are sufficient weights on the bottom to sink the net below the surface.' In 1970 over 100 of these illegal nets were removed by water bailiffs dragging the river at nights.

Coracles are also found on the Tywi, where twelve are licensed to fish for salmon in the river below the town. In the eighteen-sixties, according to the *Commissioners Report on the Salmon Fisheries,* no fewer than '400 men . . . supported themselves on the salmon and sewin fisheries'.[32] The report continues: 'To a poacher (a coracle) is invaluable . . . The coracle man is often lawless and always aggressive, he poaches private waters for years and claims a prescriptive right; he uses violence if he is very strong, he threatens if his opponent be not so much weaker than himself as to make violence unsafe . . . working without noise and at night and scarcely visible, they are difficult to detect, and if detected, almost impossible to capture, for a few strokes of the paddle will always place the river between the poacher and his pursuer'.

The authorities, even in the eighteen-sixties, were critical of the coracle as a fishing craft in non-tidal waters, for although 'in tidal waters they are a perfectly fair and legitimate engine . . . in the fresh and non-confined portions of the rivers, they are very destructive . . . if a fish shows himself in the day, his capture that night is nearly certain. Perfectly portable, the coracles are put in the river at the end of the pool, containing the fish, and it is swept again and again, if necessary until he is caught'.[33] The twelve pairs in use today between Carmarthen Bridge and the sea represent a considerable reduction in numbers since the nineteen-twenties, when in 1929, 25 pairs were licensed to fish from coracles. By 1935 the numbers had declined to 13 pairs 'under stringent regulations and subject to a licence fee of four guineas'.[34]

Coracles on River Tywi at Carmarthen, circa 1933

As on the Teifi an attempt was made to abolish coracle fishing on the Tywi in the nineteen-thirties, but by 1938 'the Ministry appears now to be willing to tolerate netting as long as the number of coracles does not exceed in future, a dozen pairs'.[34] This was the position in 1971 when twelve coracle licences were still allowed.

On the Taf, a short swift river that flows into Carmarthen Bay near the village of Laugharne, two licensees are still allowed to fish from coracles. The fishermen based on the village of Lower St Clears are part-time workers and operate at night during the months of June and July only. In the early nineteenth century, the Taf fishermen were regarded as very efficient and one observer 'saw for the first time, those feats of dexterity which are required in the management of such a capricious vessel', despite the fact that he was familiar with other coraclemen 'of the Towey and other rivers'.[35]

The numbers of salmon and sewin caught by west Wales coraclemen in 1970 and 1971 are shown in Table 8 *(see below).*

TABLE 8

RIVER	LICENSEES	LICENCE FEE	SALMON		SEWIN	
			Number	Weight (lb)	Number	Weight (lb)
1970						
Teifi	5	£15.37½	33	417	160	563¾
Tywi	12	£12.00	225	1789¾	992	2829½
Taf	2	£12.00	39	329¼	.104	238½
1971						
Teifi	6	£15.37½	23	222	146	473
Tywi	12	£12.00				
Taf	2	£12.00	188	1653	892	2661

THE CORACLE NET[36]

The coracle net used by west Wales fishermen at the present time is a movable drag net unlike any other net used in British rivers. The coracle nets used on the River Tywi are especially complex, for the weights attached to the foot-rope of a net are evenly distributed to an intricate, pre-determined pattern, according to the state of the tide and the flow of water at any one time. Considerable knowledge and experience of local conditions are vitally important in the correct setting of the Tywi net. Teifi nets on the other hand are much simpler and no rigid allowance is made for the amount of water in the river.

Basically, the coracle net is a shallow bag dragged along the bottom of the river, the mouth of which is kept open by towing it between two coracles. It consists of two sheets of hemp, linen or more recently bonded nylon, joined together at the top, bottom and sides. The armouring has a large mesh, the lint a much smaller mesh, and as the lint is much deeper being three times the depth of the armouring, it billows out as a shallow bag behind the armouring when the net is towed through the water.

The Teifi coracle net, according to Fishery Regulations,[37] consists of 'a single sheet of netting measuring not more than twenty feet in length and not more than three feet and nine inches in depth, and having meshes measuring not less than two inches from knot to knot or eight inches round the four sides and having attached round its four edges, and on one or both sides, a sheet of armour measuring not more than twenty feet in length and not more than two feet and six inches in depth and having meshes measuring not less than five and one-half inches from knot to knot or twenty-two inches round the four sides'.

The Tywi coracle net, on the other hand,[38] consists of 'a single sheet of netting measuring not more than 33 feet in length and no more than 3 feet 9 inches in depth and having meshes measuring not less than one and a half inches from knot to knot or six inches round the four sides, and having attached round its four edges and on one side a sheet of armour, measuring not more than 33 feet in length and not more than 3 feet 9 inches in depth and having meshes measuring not less than five and one-half inches from knot to knot or twenty-two inches round the four sides'. There are many differences between the nets used by coracle fishermen on the two rivers, based not so much on legal requirements, but on tradition and on the customs of the fishermen. The Tywi net is a far more sophisticated instrument than the Teifi and the complex and varied names for each part of the net suggests that the Tywi net is possibly much earlier than the Teifi. On the Teifi, hemp has always been preferred for making nets, and although hemp is occasionally used on the Tywi, the Carmarthen fishermen prefer linen thread, which they regard as being much stronger than hemp. In the eighteenth century flax was grown in the Tywi valley and this provided the raw material for the net-makers of Carmarthen. There is no evidence to suggest that linen was ever used on the Teifi and the fact that the fishermen of Teifiside used imported hemp threads for their nets, again suggests a much later development. In recent years, bonded nylon has been used by Carmarthen coracle fishermen to make nets, although until recently the fishermen disliked synthetic fibre and clung to traditional materials for nets. On the Teifi, the head, reeving and foot ropes are always of horse hair, but on the Tywi cow hair is used exclusively for the lines.

THE TEIFI CORACLE NET

The Teifi coracle net is 20 feet long and no more than 3 feet 9 inches deep. The mesh of the armouring can vary from the top to the bottom of the net, with the result that the top row of meshes, stapled by the third to the head-rope, may be no more than a couple of inches wide, while the remaining meshes may be of the full legal dimensions. In use, the top row of meshes will be stretched out to be almost parallel to the head-rope, so that in effect the armouring presents a wall of fine meshes when the net is drawn through the water. At Llechryd, the bottom row of meshes and the armouring *(y fras* or *y cefen)* is considerably larger than at Cilgerran or Cenarth, because of the rocky nature of the river bottom in that particular stretch of river. About 1½lb of hemp is required to make a coracle net, but the mesh of those nets may vary according to the time of year when it is used. In March and April when a run of large salmon is expected, a lint mesh of 6 inches is used, while in May and June, to catch small salmon *(Meillion Mai)* a smaller mesh is required.

The foot-rope *(blwm-ffun)* is made of a three-ply *(teircain)* plait of horsehair which does not absorb water and this is leaded at regular intervals with lead weights. Pieces of lead cut from sheet lead into rectangles of 2 inches by ¾ inch are pressed into shape on the foot-rope. When the river is in flood it is necessary to add more weights to the net to ensure that it sinks properly in the water. The net is suspended from a head-rope *(ffenest-ffun)* which is again made of three-ply horsehair, which according to one Cilgerran fisherman 'should be thinner than my little finger'. This is connected to the net at intervals of three or four meshes. In Teifi nets, no corks are attached to the coracle net. The stapling line *(traill-ffun)* is attached to eleven or more horn rings, threaded and running on the end rope. To make these rings, a cow horn is boiled until soft and moulded on a piece of wood. The horn is then sawn into ¼ inch rings. At one end the stapling line is made fast to a non-running horn ring lashed on to the head-rope; the outer end is tied to the last of the running rings at the opposite end of the net. To the same ring is attached one end of a long reeving line *(ffun fowr)* of which the other end is free. The reeving line is considerably thicker than the other lines in the net, 'three times as thick as the small finger' and is usually plaited from six hanks of horse hair. The horn rings are attached to the head-rope at approximate intervals of 22½ inches. The net itself is about 90 meshes wide and on the Teifi it is measured by the *gwrhyd* (ie *gwr* [man] and *hyd* [length] – the distance measured by the fully outstretched arms of a man). The reeving line between the edge of the net and a coracle should be two *gwrhyd* long.

In use, the net is carried to the river bank on top of the coracle carried over the shoulders of the fisherman allocated to the right-hand side of the pair of coracles. It is the senior partner who always takes the left-hand bank, and it is his invariable duty to draw in the net when fishing. On reaching the river bank, the net is carefully arranged and the pair of coracles enter the river drifting into the flow in mid-stream. 'The coracles will be about four or five yards apart', says one fisherman.[39] The net will be between them, the left-hand fisherman holding both reeving line and head-rope, the right-hand fisherman holding the reeving line only. When a salmon strikes the net, the right-hand man *(dyn llaw dde)* throws the reeving line and the net closes as his partner pulls it in to his side of the river. As the left-hand man *(dyn llaw whith)* draws in the net, the right-hand coracle comes up behind him, grasping the gunwale of the other coracle to prevent it drifting with the flow of the water *(colli dwr,* ie 'losing water'*).* Contrary to general belief, a salmon is never seen 'jumping in the river'; if one is seen, it is certainly a kelt and not worth catching. The action of a coracle net, therefore, is that where the reeving line is thrown by one fisherman, the strings attached to the horn rings shut together along the head-line, thus bunching the armouring meshes together. In turn, the lint meshes bunch together, so that the salmon is caught in a bag formed by the lint and armouring. At the end of a trawl, the pair of fishermen land on the river bank; they heave the coracles on their shoulders and the senior partner arranges the net and throws it on top of his mate's coracle. They then walk back to the beginning of the trawl again.

THE TYWI CORACLE NET

Although the Tywi coracle net is similar in general design and method of usage to that of the Teifi, there are many differences in detail. If the Tywi net is set off balance it will not fish properly so, says one coracle fisherman,[40] 'the arithmetic must be right. Generally speaking the bottom of the armouring is set in by the third and the head of the armouring slightly less than the third. The lint is set slightly less than the half.'

In the preparation of a coracle net the lines, made of cow's tail hair, have to be prepared first. The process of preparing these is exactly the same as that adopted by Teifi coracle men, with the exception that cow hair is used on the Tywi instead of the horse hair of the Teifi. The head line *(carn ffun),* the reeving line *(traill ffun)*, the foot line *(plwm ffun ucha)* and the hand line *(llaw ffun)* are made of three-ply strands while the second lead line *(plwm ffun isha)* is a two-ply strand. The hair is cut from a cow's tail and the longer the hair strands, the better.

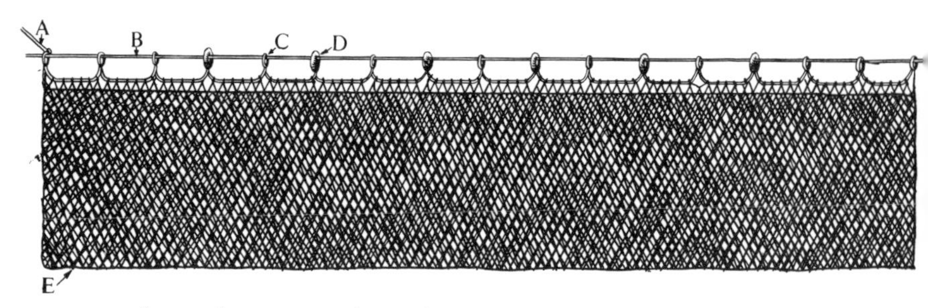

A coracle net as used on Rivers Tywi and Taf: A reeving line; B head line; C horn ring; D cork and horn ring; E foot rope

After it is thoroughly washed and picked, the unwanted hairs being disposed of, the resulting bunch of hair is placed in a shallow pan of water and covered with a heavy weight. After being steeped in water for some time, the hair is then spun into a single-ply rope, no more than an eighth of an inch in diameter. To do this a wooden rope twister *(trwc)* is used, and after about fifteen fathoms of single-ply rope has been spun, this is transferred to a second rectangular *trwc*. 'The single-ply rope', says Raymond Rees, a Carmarthen fisherman, 'is folded in half at the 30 yard mark and this end transferred to a third *trwc.* Spinning can now commence to form a two-ply rope. At the end of the original measured length, the rope is again folded and the work is completed to form the third ply. This is a three-man job and a set of *ffuniau* for a coracle net usually takes about ten days to complete.' All the ropes are made of dark brown or black hair, with the exception of the hand and reeving lines, which are invariably plaited from grey or white hairs. When fishing at night, it is important to distinguish one line from the other. The natural oils in cow hair gives the lines their water-resistant qualities and prevent undue shrinkage. The amount of twist incorporated in the construction of a line prevents stretching, and gives it its strength; although the cow hair lines are very strong, they can be broken if the net, especially the foot line, gets caught up in debris on the river bottom. This particular quality is paramount if the net is to be saved, although with the recent advent of synthetic materials, the main lines are so strong that the net has to be cut from its obstruction and the net is usually a write-off.

With all the *ffuniau* of cow hair made, the next step in the construction of a net is the preparation of the lint. The size of mesh depends on the time of season. During the early months of March and April the lint mesh size is 4½ inches made of thick linen thread, size 10/3 or Number 53 nylon. During May and June, the fish begin to run smaller and a mesh of 3¾ inches is required. This is made of 12/3 linen thread or Number 43 nylon. In July and August, small sewin appear and the lint with a 3 inch mesh is made of 18/3 linen thread or Number 33 nylon. This type of fine meshed net is known as a *Gwangrwyd*. Carmarthen coracle fishermen are unwilling to use fine-meshed nets before the run of lampreys is over towards the middle of June. A fisherman using a coarse meshed net knows when to change to a finer one because 'he feels the fish going through the coarse meshed net of the early season'. As on the Teifi, the senior partner in a pair occupies the left-hand side of the river when fishing, but a licensee may endorse up to three names on his licence. Any two members of a team may fish as long as they carry a brass tag bearing the licence number on the net. Each member of a team may possess two or three nets of each size, and one net may be used for fishing every night until it fails to catch a fish on a particular night. It is then customary to substitute another endorsee's net, until it too fails to catch a salmon.

In making a net of 3 inch mesh, the lint is braided to 340 meshes wide and 34 meshes deep. After completion it is shrunk in hot and cold water and then measured to ensure that four full meshes measure not less than 12 inches (ie 4 x 3 inches). Assuming that the four meshes measure 12 inches, then the armouring can be made. The mesh size of the armouring is dependent on the lint and it must always be ¾ inch less than

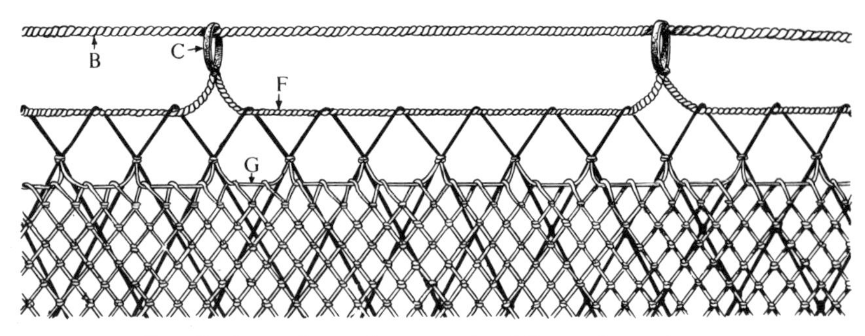

Detail of top of coracle net: B head line; C horn ring; F stapling line; G marrying

four meshes of lint. Therefore, the armouring must have a mesh of 11¼ inches from knot to knot. The length and depth of the armouring is also dependent on the lint. Since four lint meshes are equal to one armour mesh, then the lint length (320 meshes), divided by 4, equals the length of the armour, that is 80 meshes. The depth of the armouring is again determined by the lint, since the lint must be three times the depth of the armouring. It has been found by experiment that the best net is constructed 4½ meshes deep, each mesh being 11¼ inches which is within the legal maximum for depth.

The thread used to construct the armouring is usually much thicker than that used for the lint, being of 6/4 linen thread or Number 90 nylon.

The lint is now attached to the third half mesh from the head of the armouring, using a thin marrying linen thread. Four lint meshes are fixed in a continuing series of loops to each of the armour meshes until all 320 lint meshes are equally distributed in fours to the 80 armour meshes. These loops are called the *cwplins* (marryings). The next step is to complete the bag of the net by fixing it to the foot line. Rees describes the process as follows: 'To enable the armouring to hang in its correct proportions, that is by the third, it must evenly and accurately be set on the foot-rope *(plwm ffun ucha).* Two full meshes of armouring (2 x 11¼) are divided by three and that measurement (7½ inches) is the division length between each armour length on the footline. Each one of these divisions is known as a *machogyn.* The second lead line *(plwm ffun isha)* is now attached to the first lead line and the bottom lint meshes evenly distributed in fours along its length. The sides of the net are then sewn down to form the complete bag of the net.'

At this stage the finished lead line is exactly the same length as the head of the net, hung by the third. Because the net is drawn through the water in an arc, the foot line, if the net is to work properly, must be longer than the cork line. The difference in length is known as a *moilad.* 'It has been calculated', says Rees, 'that for every 20 *machogyns* on the foot-line the cork line must be 1½ *machogyns* shorter, thus on a net 80 *machogyns* long, 80/20 = 4; 4 x 1½ = the *moilad* or 6 *machogyns* difference in length.' This figure is agreed by all to be the most important measurement in the coracle net. As each *machogyn* is 7½ inches apart, the cork line must be 6 x 7½ shorter (ie 45½ inches). The cork line fully stretched is 43 feet, but because of the arc when fishing, this is reduced to less than 33 feet and well inside the legal limit.

Unlike the Teifi coracle net, the Tywi carries a line of corks and the setting of them is a complex process. Five 2 inch corks are evenly distributed between 10 cow's-horn rings on the cork line. The setting begins with fixing horn rings, followed by a cork, then a horn ring, then a cork and so on, with horn rings at the end of the line.

The armour meshes are now hung between the series of corks and horns, and always hung in the same pattern — 5, 5, 6, 6, 6, 6, 6; 6, 6, 6, 6, 6, 5, 5 = 80 meshes. It will be possible for a coracle net to be lengthened or shortened by six meshes, provided a horn ring and a cork is also added or subtracted. The ring at one end of the net is made fast to the *traill-ffun*; the other end ring is free to slide and close the net, rather like a purse string. To each of the end rings is made fast a *llaw-ffun.*

The net is now complete with the exception of being weighted with lead weights to ensure that the net fishes the bottom of the river efficiently. It is important among the Tywi fishermen that these weights are of equal thickness, dimensions and weight, and the haphazard weighting practised by Teifi fishermen is not for the Carmarthen coracle man. Weights measure 1 inch by ¾ inch thick and weigh 5 or 6 grammes. They are hand made on a pair of stone moulds 6 inches long by 4 inches wide, in which two grooves 1 inch wide are cut. The lead is melted and run into the grooves in the stone. On cooling, lead strips of even thickness are produced, and these can be easily cut to the required sizes. When leading a net, account must be taken of the flow of the river, and weights are added or subtracted accordingly. Apart from flood water, tidal flows have to be considered and each fisherman has his own particular method to enable him to decide on how much water is flowing in the river at any one time. Stone steps, bridges, quarries, even grass banks may all give a key to the flow of the river, and a net has to be weighted accordingly to the fisherman's judgement. Again, says Rees, 'The system of weighting adopted at Carmarthen is based on mathematics. Weighting always begins at the dead centre of the *plwm ffun,* that is on the fortieth *machogyn.* A single piece of lead is wound around both *plwm-ffuniau*, and working from the centre a piece of lead is fixed on the right hand side of the mesh to the right of the middle. Another lead weight is fitted on the left side of the mesh to the left of the middle of the net. In the dark, one knows exactly which side is which when fishing. Weighting is continued on both sides from the middle until the ends or *âls* are reached. As the *âls* will be close to the coracle, more weights are fixed to the ends than to the middle of the net, to allow for the extra drag.'

Accurate weighting of the net has always been the biggest problem for coracle fishermen and many a net has been discarded as being 'off balance', when in fact incorrect weighting has contributed to its failure. 'When I began fishing', continues Raymond Rees, 'it was generally accepted that one learnt the leading system parrot fashion, and it was only after careful study that a pattern appeared in the system, and praise must go to the designer of that system, whether by luck or judgement . . . for he concluded, whoever he was, that if two pieces of lead were placed side by side on the ends, then they should be followed by three single pieces. If three pieces were placed side by side then they should be followed by four double pieces followed by six single pieces of lead. As the weight of lead in the *âls* increases, then a balance must be maintained towards the middle. If one fisherman asked how much water was flowing, he would not be answered "6 to 8 inches" but by the number of pieces of lead weight on his net and always in Welsh. For example, he would be answered in ascending order any one of the following: *dau-ddau* (two twos); *tri-dau* (three twos); *pedwar-dau* (four-twos); *un-tri: pedwar-dau* (one three: four twos); *dau-tri:pump-dau* (two threes: five twos) etc.

'Each side of the net is weighted separately and by experience it has been found that the right hand side is usually a few weights lighter. If, for example, a net is weighted *un-tri; pedwar-dau* it would contain 1 x3, 4 x2, 6 x1, blank, blank and so on to the middle of the net. The right hand side would contain 1 x 3, 3 x 2, 5 x 1, blank, blank, and so on to the middle. As the flow of water increases, the blanks to the middle are filled: first forming a series of three single leads, then seven. All the blanks are taken up as in *unau* (ones) after which on every seventh a double is added as in *saith a dau* (7 x 2) and finally on every third a double weighting *tri a dau* (3 x 2). The *âls* or wings are leaded, the nomenclature of the various leadings being given as follows:

1 x 2: 3 x 1 *un a pob yn ail* to the middle of the net
2 x 2: 4 x 1 *dau a pob yn ail* "
3 x 2: 5 x 1 *tri a pob yn ail* "
4 x 2: 6 x 1 *pedwar a pob yn ail* "
1 x 3: 4 x 2: 6 x 1 - *un, tri a dim* "
2 x 3: 5 x 2: 7 x 1 - *dau tri a dim* "
3 x 3: 6 x 2: 8 x 1 - *3 times tri un a dim* "
4 x 3: 7 x 2: 9 x 1 - *4 times tri un a dim* "
1 x 4: 5 x 3: 7 x 2 - *full saith a dwy* "

2 x 4: 6 x 3: 9 x 2 - *full un a dim* to the middle of the net
3 x 4: 7 x 3: 10 x 2 - *full saith a dwy* "
4 x 4: 8 x 3: 11 x 2 - *full tri a dwy* "

'Of course the ends of the net can be made lighter by subtraction and corresponding subtraction from the middle, but that is where experience and knowledge of river conditions come in.'

Of course the setting of the Tywi net is by far the most complex of all, for as far as can be ascertained, complicated mathematical formulae never entered the calculations of coracle fishermen on other rivers. It is interesting to note that, although many Carmarthen coracle men like Raymond Rees are not Welsh speaking, their terminology and fishing nomenclature is always in the Welsh language and the persistence of fishing terms, probably from medieval Welsh, may be an indication of the antiquity of the coracle and the coracle net, on the Tywi in particular.

CORACLE TERMS USED BY FISHERMEN ON TEIFI AND TYWI

ÂL (Tywi) – the heavily leaded wings or selvage of a net.

ASEN, pl *EISAU*, *EISE* (Teifi), *ISE* (Tywi) – the ribs or laths of a coracle.

ASEN SAETHU (Tywi) – the two main cross ribs of a coracle running diagonally across the bottom.

ASTELL ORLES (ORLAIS) (Tywi) – the plank, usually of oak, at right angles to the seat, supporting it and forming a box for the catch.

BACHYN (Tywi) – a forked stick for holding down the lath framing of coracles during construction.

Y BLETH (Teifi) – plaited hazel gunwale of coracle, *cf PLETH FAWR* (Tywi).

BREST Y CORWG (Tywi) – front of coracle, *cf BLA'N Y CORWG* (Teifi).

BLWMFFUN (Teifi) – leaded foot rope of a net, *cf PLWMFFUN* (Tywi).

BWRW (Teifi) – a cast, ie a stretch of river; *PEN-BWRW* – the starting point of a cast.

CARNFFUN, GARFFUN (Tywi) – head-rope of a coracle net (*carn* [horn] + *ffun* [line]), *cf FFENEST-FFUN* (Teifi).
CLAWR (Tywi) – a net mesh gauge, *cf PREN MAGAL* (Teifi).
Y CEFEN (Teifi and Tywi) – the lint of a coracle net.
CLYFWCHWR (Tywi) – the time to begin fishing, ie twilight.
CNOCER (Teifi and Tywi) – knocker or priest for killing salmon, *cf MOLLY KNOCKER* of lower Severn (molly = Severn name for salmon), occasionally on Teifi the *cnocer* is referred to as *PREN PYSGOD* (fish stick).
CWRWGL, CWRWG, CORWG (Teifi and Tywi) – a coracle.
CWT Y CORWG (Teifi and Tywi) – the rear of a coracle.
DELLTO (Teifi) – splitting laths for coracle frame.
DREI-FFUN (Tywi) – reeving line, *cf FFUN FOWR* (Teifi).
EISE CROS (Teifi), *ISE CROS* (Tywi) – cross laths.
EISE HYD (Teifi) – longitudinal laths of a coracle, *cf ISE HIR* (Tywi).
FADDUG (Tywi) – the V-shape of meshes on head-rope and foot-rope.
FFENEST-FFUN (Teifi) – head rope of net, *cf CARN FFUN* (Tywi).
FFIOL (Tywi) – wooden baler, not used on Teifi.
FFUN, pl *FFUNIAU* (Teifi and Tywi) – the cords or ropes of a net.
FFUNEN (Teifi) – fishing rod.
FFUN FOWR (Teifi) – reeving line, *cf DREI-FFUN* (Tywi).
Y FRÂS (Teifi) – armouring of net, *cf RHWYD RÔTH* (Tywi).
GAFEL (Teifi) – the claw *(gafael)* at the top of a Teifi coracle paddle.
GASEG (Teifi) – shaving horse used for shaping laths, not often used by Tywi coracle builders.
Y GORON (Tywi) – the spot where the two diagonal laths *(eise saethu)* cross one another.
GWANGRWYD (Tywi) – a fine meshed net of about 3 inches bar for use in catching small salmon and sewin, usually up to 1lb in weight; mainly used at the end of the season.
GWAS (Teifi) – endorsee of coracle licence.
GWDEN (YR WDEN) (Teifi) – a twisted withy, hazel or oak

sapling used for carrying coracle, *cf STRAPEN GWDDWG* (Tywi).

GWE FRÂS (Teifi) – armouring of net.

GWRAGEN (Tywi) – a lath loop around the front of the gunwale of a Tywi coracle for the protection of the outside.

GWRHYD (Teifi mainly) – a measure of the length of a coracle net. *GWR* (man) + *HYD* (length), ie a measure of a man's outstretched arms.

HALA CWRWG (Teifi) – to be a coracle fisherman, literally 'to send a coracle'.

HELINGO (Teifi) – the process of covering a coracle.

LLAW FFUN (Tywi) – stapling line (handline).

LLYGAD (Tywi) – mesh, *cf MAGAL* (Tywi and Teifi).

MACHOGYN (Tywi) – the gap between meshes of a net or *PLWM-FFUN;* for every 20 *machogyn* on the lead line, the cork line must be 1½ *machogyns* shorter.

MAGAL, pl *MAGLE* (Tywi and Teifi) – mesh of net.

MAGLE BACH (Teifi) – lint of coracle net.

MAGLE MOWR (Teifi) – armour of a coracle net.

MOELYD NOL (Tywi) – to go back to the beginning of a trawl or pool by paddling the coracle rather than carrying it.

MOILAD (Tywi) – the difference between length of cork line and lead line; too much *moilad* will mean that each mesh is elongated; too little *moilad* means that the meshes are stretched horizontally.

OFEL (Tywi) – frame of a coracle.

ORAGE (Teifi) – frame of a coracle.

PLETH FACH (Teifi) – withy plait that runs from the gunwale and along the bottom of a Teifi coracle to give added strength at the back.

PLETH FAWR (Tywi) – wattled gunwale, *cf Y BLETH* or *Y BLETH DOP* (Teifi).

PITCHO CORWG (Teifi and Tywi) – to put pitch on a coracle; referred to occasionally as *RHOI COFOR* (putting on a cover)

PREN MAGAL (Teifi) – a net mesh gauge, *cf CLAWR* (Tywi).

PLWM-FFUN (Tywi) – leaded foot rope of a coracle net.
Plwm-ffun ucha – upper lead line attached to lint –

three-ply cow-hair ropes; *Plwm-ffun isha* – lower lead line attached to armouring every third mesh – two-ply cow-hair rope.

PWYLLO'R FFUNIAU (a) arranging the rope and nets in a coracle (Tywi); (b) the bunching of the net meshes together when a fish is caught (Teifi).

RHAWN – horse hair (Teifi) or cow hair (Tywi) used in making coracle net ropes.

RHWYD FÂN (Tywi) – lint of a coracle net.

RHWYD RÔTH (Tywi) – armouring of a coracle net.

SDOL (Teifi) – ie stool – coracle seat, *cf SÉT* (Tywi).

STRAPEN CNOCER (Tywi) – strap to hold knocker in place on seat.

STRAPEN FFIOL (Tywi) – strap holding baler in place.

STRAPEN GWDDWG (Tywi) – the ash or leather carrying strap, *cf GWDEN* (Teifi), also *STRAPEN CWRWG.*

TEIRCAIN (ie *TRI* [three] + *CAIN* [hank]) – the three-ply horse-hair (Teifi) or cow-hair (Tywi) rope of the *ffenest-ffun* (headrope) and *plwm-ffun* (lead line).

TRAILL-FFUN (Teifi) – stapling line carrying horn rings.

TRYC, TRWC (Tywi) – a rope twister, somewhat similar to that used in making straw rope for making the *ffuniau* (ropes) of a coracle net.

WAL (Tywi) – lint of a net, *cf Y CEFEN, RHWYD FAWR* (Teifi).

WECHCAIN (ie *CHWECH* [six] + *CAIN* [hank]) – six-ply of horse hair (Teifi) or cow hair (Tywi) of the *ffun-fowr* (reeving line).

YSGAR (Tywi) – half a mesh.

YSGAR Y GATH – the lowest line of meshes in a net, usually on the armouring; this must be large enough for a salmon to get through to the lint beyond.

THE TEIFI CORACLE

The Teifi coracle, used by generations of fishermen at Cenarth, Aber-cuch, Llechryd and Cilgerran, varies little in shape and design from one part of the river to the other. According to Hornell,[41] 'the Teifi coracle is characteristically short and of

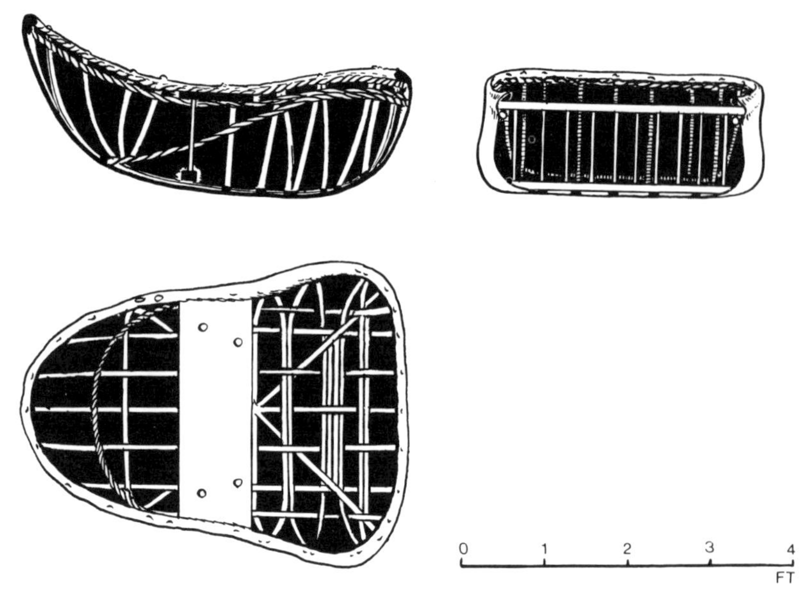

A Teifi coracle

squat ungainly shape; in plan broad and with very little horizontal curve at the fore end; nearly semicircular in plan at the after end. At the insertion of the seat, placed about mid-length, the gunwale is pinched in at each side, giving the appearance of a slight waist between the forward and after sections. At the fore end and along the sides to a point just behind the seat, the coracle shows a slight degree of tumble-home, whereby the bottom view appears broader than the face plan and has no midships constriction, its outline being bluntly triangular, with all the angles well rounded. The apex, more rounded than the other angles, represents the stern. The gunwale sheers slightly towards the fore end, more emphatically towards the after end. The bottom is flat except for the last 12-15 inches, where it curves up gradually to the extremity of the stern. To anyone unfamiliar with these coracles, the narrowed and curved-up stern would seem to be the fore-end, whereas the wide and deep forward end would certainly be considered as the "stern" '.

The framework of the coracle *(yr orage)* consists of seven longitudinal laths *(eisau,* sing *asen* or *eise hyd)* interwoven at right angles by seven transverse laths all spaced four or five inches apart. Crossing each other in front of the seat are a pair of diagonal laths *(eisau saethu,* sing *Asen saethu)* that are also interwoven with the longitudinal and transverse laths. As only one transverse lath is found behind the seat, this part of the coracle is strengthend by a semi-circular plait of hazel withies. Behind this plait the frame of the coracle is bent upwards to the stern of the vessel, and the position of the hazel plait marks the rear limit of the flat floor of the craft. To construct the coracle frame, willow branches are selected and cut; those for the longitudinal laths will be 7 feet 6 inches or 8 feet long and those for the transverse laths 5 feet or 6 feet long.[42] These are cut in the autumn or winter when they are not full of sap, the best quality willow being approximately seven-year-old pollard willow. In the past the demand for willow, which grows profusely in the Teifi valley, was considerable and willow trees were lopped at regular intervals. Today, however, with the disappearance of so many willow industries, harvesting is no longer a regular occurrence, and coracle builders are finding increasing difficulty in obtaining the correct raw material for framing coracles. With a billhook, each willow rod is split in half, so that the two sections are of equal thickness throughout their lengths. Each piece of timber is then placed in a shaving horse and smoothed with a two-handed draw-knife and spoke shave. The coracle builder's shaving horse is a low bench which the craftsman sits astride, pressing the clamping pedal with his foot, so that the lath, held fast by the clamp block, is held in the correct position for shaping. Before use, the laths are soaked in hot water or wrapped round with cloth dipped in boiling water so that they are more pliable. Seven laths are then laid on the floor, or on a wooden board, spaced 4 or 5 inches apart, and ten of the shorter laths are interlaced with them at right angles. The first group of transverse laths are laid close together in three pairs so as to provide added strength under the feet of the coracle man, so that, in reality, there are only seven transverse ribs in the coracle frame. The two long laths are arranged diagonally across the front of the laths and interlaced with the others. Heavy stone weights are placed at the intersection of the laths and the sides are bent upwards. A deal seat, about 36 inches long and 11 inches wide, is placed in position and the ends of the second and third transverse laths (counting from the back) are bent up and passed through slots made near the end of the deal plank. The ends of the two main longitudinal laths are next bent up and secured in the correct position by strings stretched between them. They serve as guides in the shaping of the hazel gunwale *(y blêth dop)* that is woven from a number of 9 foot lengths of hazel twigs. The weaving usually starts from the left-hand side near the seat, and as the craftsman continues weaving towards the front of the coracle, the laths are inserted and carefully woven into the gunwale. The process is continued until the whole of the fore-part *(part bla'n)* of the coracle frame, which is almost a half-circle, is completed. At the back *(cwt y cwrwg)* one plait is bent down to form a strengthening band around the after end of the bottom, and a second is carried around continuously along the top, with the laths inserted into it. A third plait is then wattled on, although in recently made examples, a third plait is not used. When it is constructed, its withies are stronger than those in the other two, as this plait forms the margin of the gunwale and has to stand heavy usage. It passes over the ends of the seat, which are thus sunk about 1½ inches below the top of the gunwale.[43] Nine or ten wooden stanchions, the lower ends sunk into a wooden bar turning transversely across the back of the coracle, are required to give strength to the seat.

The frame of the coracle is then complete, a coat of creosote is added as soon as it is dry, and approximately 4½ yards of unbleached calico, usually obtained from a Newcastle Emlyn draper, is stretched tightly over the frame and over the gunwale. The Teifi coracle differs from the Tywi in that the gunwale is covered with calico. The process of covering is known as *helingo* and the calico is sewn into place with twine or thin wire passed through the wattled gunwale. Finally a coat of pitch boiled together with linseed oil and possibly lard or tallow is applied hot to the outer surface of the coracle.[44] A dry day is usually chosen for this process of pitching the coracle *(pitsho cwrwg)* usually carried out by women. Finally, a pair of holes is made at each end of the seat in order to fit the carrying strap of twisted oak or elm *(yr wden* or *gwden).* The normal method of shaping the *wden* is to choose an oak sapling which is twisted, but to avoid cutting to within about 4 inches of the root. It is left hand grips the loom about 15 inches lower down'.[45] The cut as near the root as possible. One end of the *wden* is passed down through the forward hole in a pair in the seat, and brought up through the after hole, thus locking it securely. The other end is treated similarly. Occasionally, a hole is bored just below the gunwale at the back of the coracle, so as to drain out the water when the vessel is carried ashore.

The paddle used on the Teifi is long and thin, made of either larch, ash or elm and is 50 inches long. Most coracle men prefer larch paddles because they are lighter than those of ash or elm; an important consideration when they have to be paddled with one hand. The blade is 16 inches long, with sharp shoulders, and while one side is flat, the other is slightly rounded. The flattened top of the paddle ends in a claw which can be slipped under and thereby engage the basal bar of the seat support when the coracle is hoisted upon the owner's shoulders. The loom (that is the shank or handle of the paddle) 'rests on his right shoulder, and thus adjusted the pressure of the carrying rope across his chest is considerably reduced. The claw also affords a useful grip for the fingers of the right hand when paddling straight ahead with the paddle used over the bow; the left hand grips the loom about 15 inches lower down'.[45] The loom of the paddle is cylindrical in cross-section near its junction with the blade, but it flattens, and for the last 24 inches of its length it is nearly rectangular in cross-section to end in the claw. In paddling a Teifi coracle, a figure of eight motion is described in the water, and the paddle is kept continuously in the stream. When fishing, paddling is done with either the left or right hand, depending on whether the coracle is on the left- or

A pair of Teifi coracles at Cenarth, circa 1930

right-hand side of the river. The loom is gripped with one hand well below the middle and the claw of the paddle rests against the shoulder. When not netting, the coracle is propelled straight ahead by paddling over the fore end 'either by figure of eight stroke or by a scooping motion. In both cases, the paddle is gripped with the two hands, one gripping the claw at the top and the other holding the loom some distance down'.

Although all the Teifi coracles show remarkable uniformity of construction, the size and proportions can vary considerably according to the preferences of the individual fisherman. A tall and heavy man requires a larger coracle than a smaller man, who may require a lighter coracle for reasons of easy porterage. The following *(see page 139)* are examples of the dimensions of Teifi coracles: Coracle A, now at the Welsh Folk Museum (Accession Number 04.199) was made and used at Llechryd c 1890.

Coracle B, also at the Welsh Folk Museum (Accession Number 51.252), was built by the Cenarth coracle builder, John Thomas, Bronteifi, in 1951.
Coracle C, was measured by Hornell in the nineteen-thirties and belonged to Alfred E. Griffiths, Cenarth.

	A	B	C
Overall length (along gunwale)	54in	58in	54in
Maximum width (near fore-end)	41½in	39in	39in
Width (at seat)	34½in	34in	34in
Depth to underside of seat	12½in	12in	13in
Height from ground Front	18½in	18in	16in
Height from ground Back	19¾in	23in	18in
Width of seat	10in	9in	—
Weight	28lb	32lb	29lb

In the nineteen-thirties a coracle would cost about £2 and there were two specialised coracle builders at Cenarth and another at Llechryd.

Coracles have been known on the Teifi from at least the last quarter of the eighteenth century, and many travellers to West Wales wrote of the coracles they saw at Cenarth, Cilgerran and Llechryd. H. P. Wyndham in 1781[46] for example, in addition to describing Tywi coracles in detail, mentions them in use at Cilgerran, where they were used for ferrying people across the river in addition to fishing. 'The dexterity of the natives', he says, 'who fish in these coracles is amazing, though it frequently happens to the most expert, that a large fish will pull both the boat and the man under water.' Malkin, some twenty years later,[47] describes the Teifi coracle in considerable detail. 'They are made with very strong basket-work and covered with hides or coarse canvas, with a thick coating of pitch. Their shape resembles the section of a walnut shell, their length is generally five feet and their breadth seldom less than four. They contain but one person and it is entertaining to observe the mode in which they are managed. The dextrous navigator sits precisely in the middle, and it is no trifling part of his care to keep his

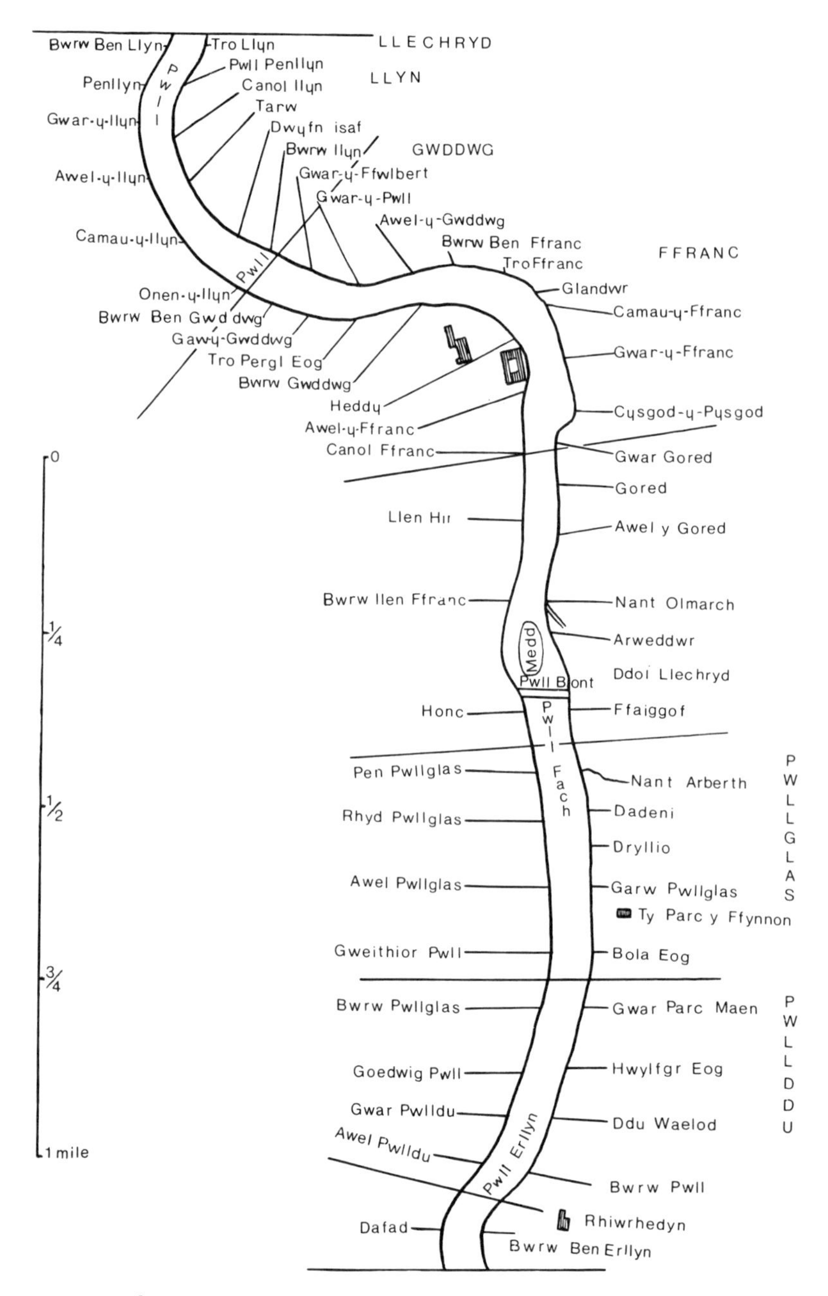

The extent of the salmon fishing grounds of the Llechryd coracle fishermen

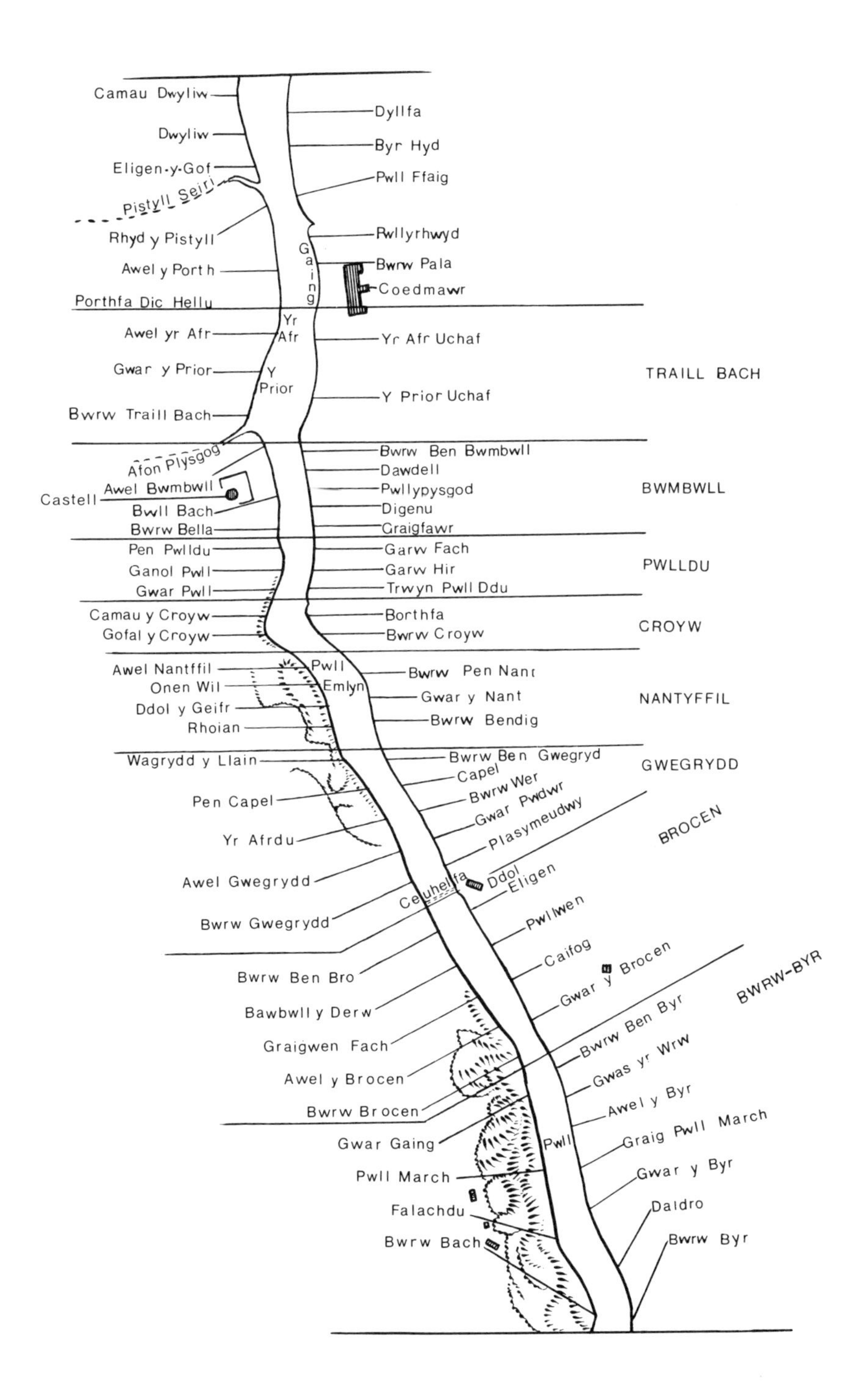

Camau Dwyliw
Dwyliw
Eligen-y-Gof
Pistyll Seiri
Rhyd y Pistyll
Awel y Porth
Porthfa Dic Hellu
Dyllfa
Byr Hyd
Pwll Ffaig
Rwllyrhwyd
Bwrw Pala
Coedmawr
Gaing
Awel yr Afr
Yr Afr
Yr Afr Uchaf
Gwar y Prior
Y Prior
Y Prior Uchaf
Bwrw Traill Bach
TRAILL BACH
Afon Plysgog
Awel Bwmbwll
Castell
Bwll Bach
Bwrw Bella
Bwrw Ben Bwmbwll
Dawdell
Pwllypysgod
Digenu
Graigfawr
BWMBWLL
Pen Pwlldu
Ganol Pwll
Gwar Pwll
Garw Fach
Garw Hir
Trwyn Pwll Ddu
PWLLDU
Camau y Croyw
Gofal y Croyw
Borthfa
Bwrw Croyw
CROYW
Awel Nantffil
Onen Wil
Ddol y Geifr
Rhoian
Pwll Emlyn
Bwrw Pen Nant
Gwar y Nant
Bwrw Bendig
NANTYFFIL
Wagrydd y Llain
Bwrw Ben Gwegryd
Capel
GWEGRYDD
Pen Capel
Bwrw Wer
Gwar Pwdwr
Yr Afrdu
Plasymeudwy
BROCEN
Awel Gwegrydd
Celuhelfa
Ddol
Eligen
Bwrw Gwegrydd
Pwllwen
Caifog
Bwrw Ben Bro
Gwar y Brocen
Bawbwll y Derw
BWRW-BYR
Graigwen Fach
Bwrw Ben Byr
Gwas yr Wrw
Awel y Brocen
Awel y Byr
Bwrw Brocen
Gwar Gaing
Pwll
Graig Pwll March
Pwll March
Gwar y Byr
Falachdu
Daldro
Bwrw Bach
Bwrw Byr

just balance. The instrument with which he makes his way is a paddle. One end rests upon his shoulder, and the other is employed by the right hand in making a stroke alternately on each side. The left hand is employed in conducting the net, and he holds the line between his teeth . . . They are now applied only to the purpose of fishing.'

By 1861 the Teifi was regarded as 'the headquarters of coracle fishing', and the *Report of the Commissioners appointed to inquire into Salmon Fisheries* stated that fishing from coracles had been 'developed to the utmost . . . The feelings of the coracle fishers are strongly antagonistic both to the fishermen at the mouth' (ie the seine netsmen of the estuary) and to the upper proprietors. They form a numerous class, bound together by a strong *esprit de corps*, and from long and undisturbed enjoyment of their peculiar mode of fishing, have come to look upon the river almost as their own, and to regard with extreme jealousy any sign of interference with what they consider their rights. In the deep and narrow water near the Cilgerran slate quarries, several pairs of coracles are sometimes so arranged as 'almost entirely to close the passage against the fish'.

On the Teifi when coracle fishing was fairly unrestricted, as it was during the nineteenth century, the river was divided into four sections, the fishermen from one of four villages having the sole right to fish in those sections of the river. The four villages concerned were Cilgerran, Llechryd, Aber-cuch and Cenarth. Each of these stretches of river was divided into a number of sections each section being termed a *bwrw* (cast).[48] Each *bwrw* was divided into three parts, each of which was called a *traill* (trawl). At Cilgerran, for example, the principal trawl *(y draill)* was that side of the river usually, but not always, nearest the village; the second trawl *(yr ail draill)* signified the middle of the river and *yr hawel* or *tu'r dre* signified a third trawl, on the opposite side of the river from the principal trawl. 'If the principal trawl was on the Cardiganshire third of the river, the third trawl would occupy the third part of the river nearest to Cilgerran and was called *Tu'r dre* or next to the town, but if on the contrary, the principal trawl was along the

Pembrokeshire third of the river, the third trawl would be . . . on the Cardiganshire side and would in that case be called *yr Hawel*.'[49]

According to local tradition, each of the eight casts belonging to Cilgerran coracle men had its own characteristics, expressed in a doggerel verse passed down over many generations to the present day:

> *Bwrw byr hyfryd – Brocen ddryslyd*
> (Lovely Bwrw byr – complex Brocen)
>
> *Gwegrydd lana – Nantyffil lwma*
> (Cleanest Gwegrydd – Poorest Nantyffil)
>
> *Crow'n rhoddi – Pwll du'n pallu*
> ([when] Crow gives – Pwlldu refuses)
>
> *Bwmbwll yn hela – Draill fach yn dala*
> (when one hunts at Bwmbwll – you may catch at
> Draill fach)

Each of the main casts had minor casts attached. Bwrw Byr had the minor casts of Pwll March and Gaing attached; Brocen had Bawbwll and Graigwen; Gwegrydd had Capel and Bwrw Oer; Nantyffil had Pwll Emlyn; Pwll du had Garw Bach and Garw Hir; Bwmbwll had Pwll-y-pysgod; Traill bach had Afr and Prior, while Bwrw Crow had no minor casts. In addition to the eight main casts and their subdivisions and accompanying minor casts, there were also eighteen other minor casts that did not belong to the principal ones. They were Pwll-y-Rhwyd (Netpool), Gaing, Pwll March, Bawbwll, Graigwen, Porthfa Lodge, Capel, Bwrw Oer, Pwll Emlyn, Pwll Elai, Garw Bach, Garw hir, Pwll-y-pysgod, Yr Afr, Y Prior, Traill y Bridill and Traill Silian (spawning shore) and Pwll Trewindsor.[50]

'In or about the month of April', says the Cilgerran historian,[51] 'the town crier used to convene by a public cry, a meeting of all the fishermen for the season . . . ' Formerly none were permitted

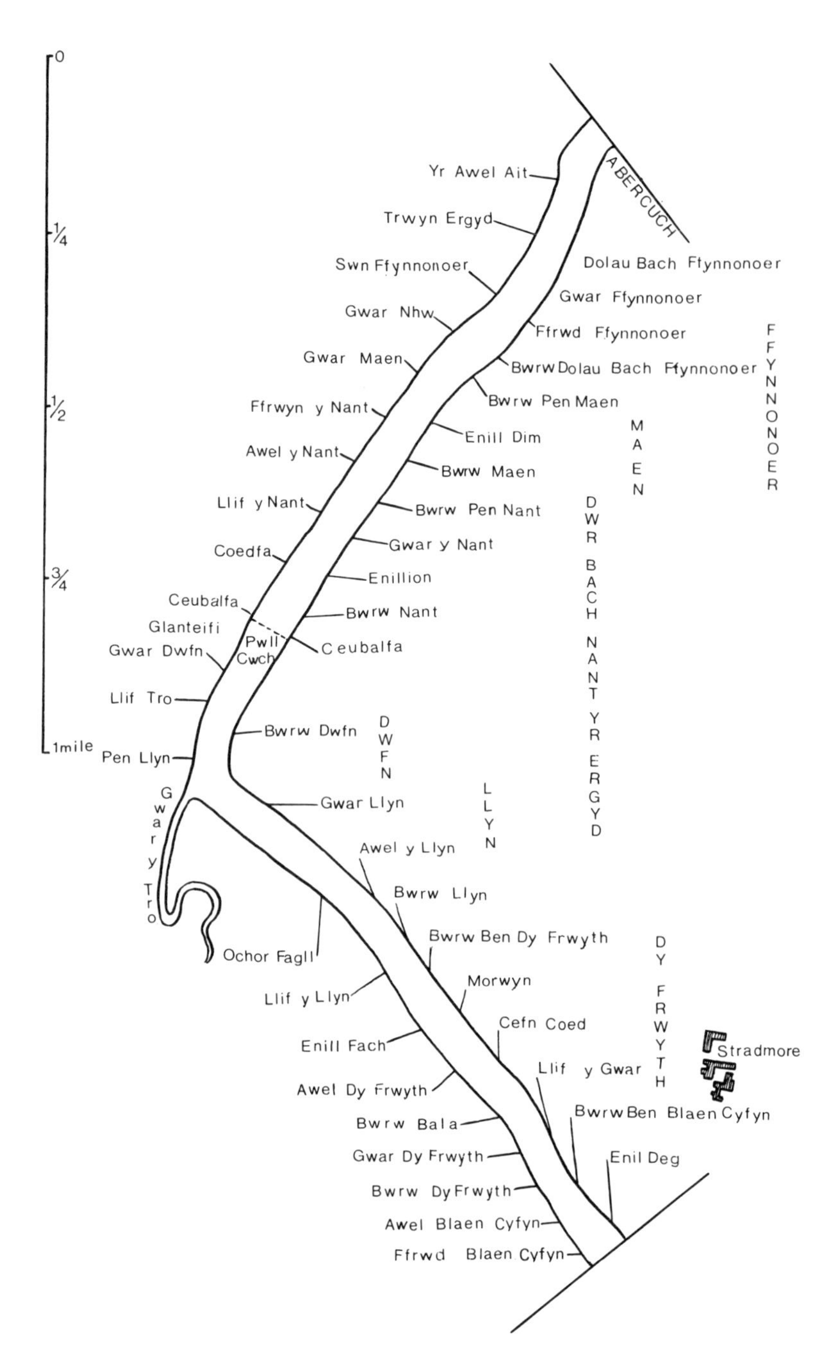

The extent of the salmon fishing grounds of the Aber-cuch and Cenarth coracle fishermen

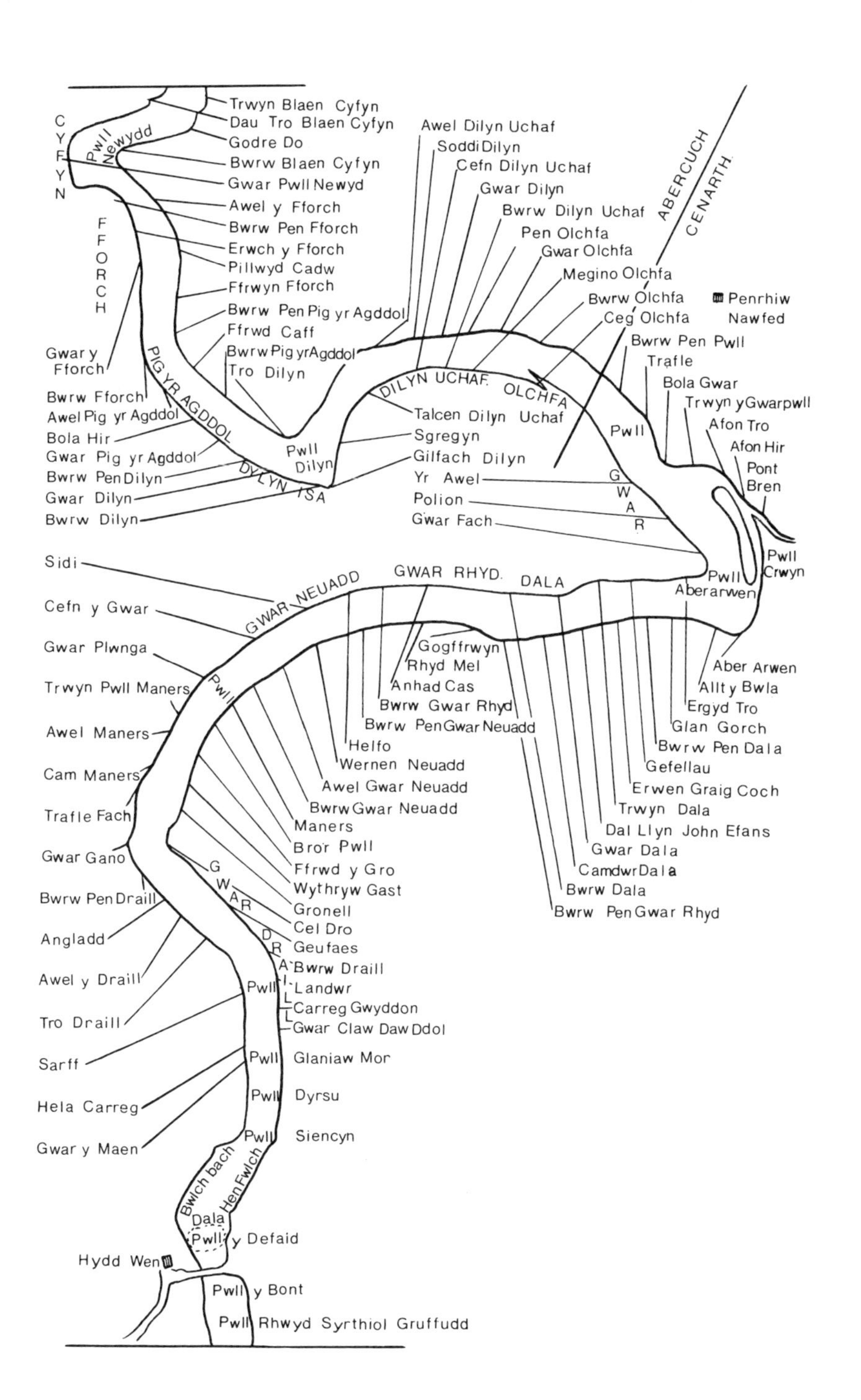

CYFYN
Pwll Newydd
FFORCH
Trwyn Blaen Cyfyn
Dau Tro Blaen Cyfyn
Godre Do
Bwrw Blaen Cyfyn
Gwar Pwll Newyd
Awel y Fforch
Bwrw Pen Fforch
Erwch y Fforch
Pillwyd Cadw
Ffrwyn Fforch
Bwrw Pen Pig yr Agddol
Ffrwd Caff
Bwrw Pig yr Agddol
Tro Dilyn
Gwar y Fforch
Bwrw Fforch
Awel Pig yr Agddol
Bola Hir
Gwar Pig yr Agddol
Bwrw Pen Dilyn
Gwar Dilyn
Bwrw Dilyn
PIG YR AGDDOL
DYLYN ISA
Pwll Dilyn
Awel Dilyn Uchaf
Soddi Dilyn
Cefn Dilyn Uchaf
Gwar Dilyn
Bwrw Dilyn Uchaf
Pen Olchfa
Gwar Olchfa
Megino Olchfa
Bwrw Olchfa
Ceg Olchfa
ABERCUCH
CENARTH
Penrhiw Nawfed
Bwrw Pen Pwll
Trafle
Bola Gwar
Trwyn y Gwarpwll
Afon Tro
Afon Hir
Pont Bren
DILYN UCHAF
OLCHFA
Talcen Dilyn Uchaf
Sgregyn
Gilfach Dilyn
Yr Awel
Polion
Gwar Fach
Pwll
GWAR
Pwll Aberarwen
Pwll Crwyn
Sidi
Cefn y Gwar
Gwar Plwnga
Trwyn Pwll Maners
Awel Maners
Cam Maners
Trafle Fach
Gwar Gano
Bwrw Pen Draill
Angladd
Awel y Draill
Tro Draill
Sarff
Hela Carreg
Gwar y Maen
GWAR NEUADD
GWAR RHYD.
DALA
Pwll
Gogffrwyn
Rhyd Mel
Anhad Cas
Bwrw Gwar Rhyd
Bwrw Pen Gwar Neuadd
Helfo
Wernen Neuadd
Awel Gwar Neuadd
Bwrw Gwar Neuadd
Maners
Bro'r Pwll
Ffrwd y Gro
Wythryw Gast
Gronell
Cel Dro
Geufaes
Bwrw Draill
Landwr
Carreg Gwyddon
Gwar Claw Daw Ddol
GWAR DRAILL
Pwll
Pwll Glaniaw Mor
Pwll Dyrsu
Pwll Siencyn
Aber Arwen
Allt y Bwla
Ergyd Tro
Glan Gorch
Bwrw Pen Dala
Gefellau
Erwen Graig Coch
Trwyn Dala
Dal Llyn John Efans
Gwar Dala
Camdwr Dala
Bwrw Dala
Bwrw Pen Gwar Rhyd
Bwlch bach
Hen Fwlch
Dala
Pwll y Defaid
Hydd Wen
Pwll y Bont
Pwll Rhwyd Syrthiol Gruffudd

to fish in that part of the Teivi which borders on the confines of this parish, save those who had previously been admitted burgesses of the ancient borough of Cilgerran, so that the river being entirely monopolised by them, strangers were effectively excluded from participating in the fishery. At the annual April meeting of the fishermen the turns or casts for the first night of the fishing season were allocated. For this purpose, slips of paper on which were written the names of the eight principal *bwrws* and also of the subdivisions of each of such casts, were deposited in a hat, from which every fisherman in his turn picked out a slip and whatever name or position might be written on that slip would be his station during that first night. This arrangement, of course, only held good for the first night, for he that would have the best chance on the first trawl on the upper cast on the first night, would be the last on the next cast on the following night, and so on until he had gone through all the subdivisions in each *bwrw*.' If a fisherman should absent himself from his station on any one night, no one could take his place, for the allocation of casts was very rigidly enforced at all times. The starting point of every station was termed *Pen bwrw* (the head of the cast), where the coracles were placed in order of precedence. 'The two leading coracles were required to be with their keels on the ground, in the same position as when on the water, with the paddle resting on the seat. If this rule was not adhered to, the owners of the leading coracles were deprived of their trawl.'

The rigid rules of precedence and privilege were undoubtedly framed by the fishermen themselves and were invaluable in that disputes were avoided. They were oral laws passed down from father to son and were practised with little variation by the coracle men of Llechryd, Aber-cuch and Cenarth as well as those of Cilgerran. Before the eighteen-sixties the close season for salmon fishing was largely a matter of unwritten law rather than of legislation and the season extended from August to February. One fisherman in each village was responsible for locking up the coracles during the close season.[52] In the eighteen-sixties, appreciable quantities of salmon were salted and smoked, and some fresh salmon was 'sent to London and those that are fit are sold to Londoners, and the rest are sent to France'.[53] In addition to coracle fishing at Cenarth, Llechryd, Aber-cuch and Cilgerran, coracle fishing was also practised up-river at Llandysul where coracle nets were in constant use for a few years. According to the Commissioners' report, coracle fishing had only been recently introduced above Cenarth Falls, and the evidence suggests that after 1863 when licences were introduced, coracle fishing at Llandysul disappeared.

In the coracle fishing villages below Cenarth Falls, fishermen flourished despite the introduction of a five shilling licence fee.[54] Nevertheless, the eighteen-sixties saw the end of the rigid rules of precedence that had been practised on the river. At Cilgerran, for example, the rules were disregarded ' . . . and recently the person who left the first coracle at a station during the day was entitled to precedence in trawling the following night, the river from that station to the next; but now before a person has any certainty of a draw, he must need place himself at the starting point and there remain with his coracle from the morning till the evening sets in, and the darkness enables him to spread his net to advantage; and even this patient watching does not always now secure a first position. In fact there is no regulation whatever adhered to; everybody scrambles for the first chance; and everybody spreads his net wherever he thinks it likely to obtain a fish'.[55]

Although the rigid rules of precedence had disappeared from Cilgerran by the eighteen-sixties, vestiges still remained until the nineteen-thirties. Cilgerran coracles that could fish between Llechryd and Cardigan Bridge fished well-defined pools – Y Gwddwg, Llyn, Pwll byr, Y Frocen, Gwaun Morgan, Y Cafan, Nantyffil, Y Crown, Bwmbwll, Bwrrwyd, March, Pen Pwll, Pysgod, Chwarel Aubrey, Y Gwter, Yr Afar, Y Prior, Ben Jubilee. The first pair of coracles at a particular pool could enter the water first. This first pair were known as *yr ergyd*. It was then followed by *yr ail draill* that could enter the water as soon as *yr ergyd* had drifted 50 yards from the start of the trawl. This was then followed by *y trydydd* (the third); but the first pair

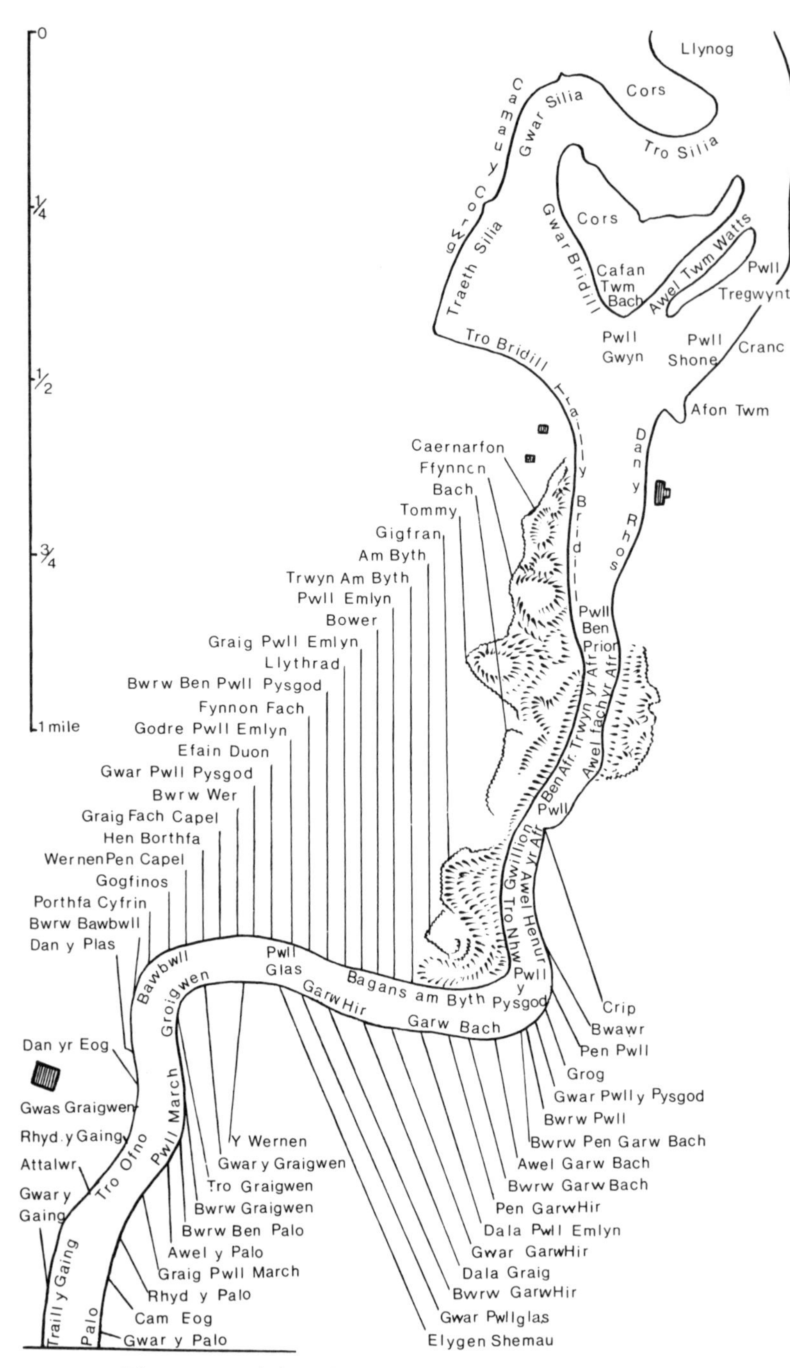

The extent of the salmon fishing grounds of the Cilgerran coracle fishermen

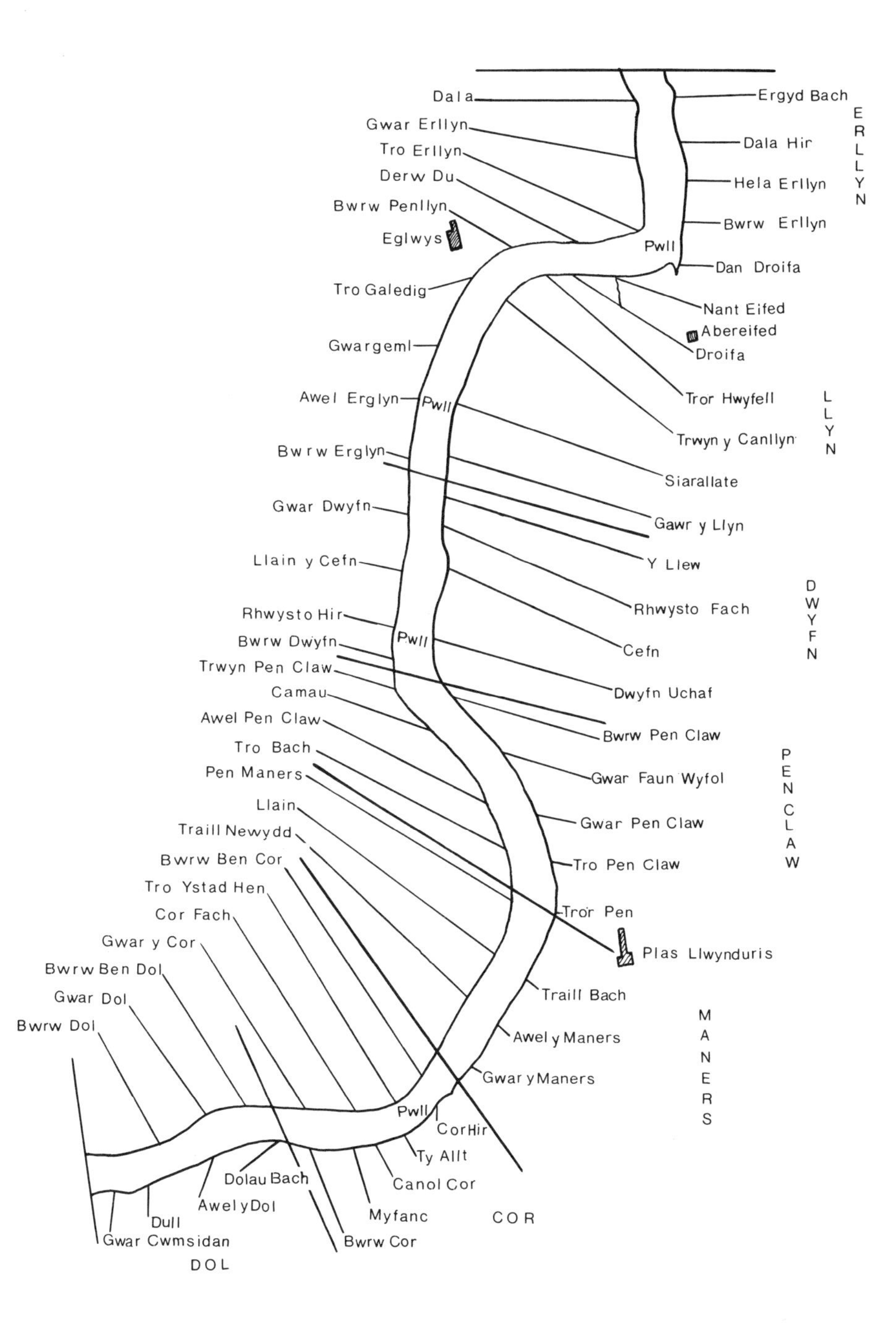
Dala
Gwar Erllyn
Tro Erllyn
Derw Du
Bwrw Penllyn
Eglwys
Tro Galedig
Gwargeml
Awel Erglyn
Bwrw Erglyn
Gwar Dwyfn
Llain y Cefn
Rhwysto Hir
Bwrw Dwyfn
Trwyn Pen Claw
Camau
Awel Pen Claw
Tro Bach
Pen Maners
Llain
Traill Newydd
Bwrw Ben Cor
Tro Ystad Hen
Cor Fach
Gwar y Cor
Bwrw Ben Dol
Gwar Dol
Bwrw Dol
Ergyd Bach
Dala Hir
Hela Erllyn
Bwrw Erllyn
Pwll
Dan Droifa
Nant Eifed
Abereifed
Droifa
Tror Hwyfell
Trwyn y Canllyn
Pwll
Siarallate
Gawr y Llyn
Y Llew
Rhwysto Fach
Cefn
Pwll
Dwyfn Uchaf
Bwrw Pen Claw
Gwar Faun Wyfol
Gwar Pen Claw
Tro Pen Claw
Tror Pen
Plas Llwynduris
Traill Bach
Awel y Maners
Gwar y Maners
Pwll
Cor Hir
Ty Allt
Canol Cor
Myfanc
Bwrw Cor
Dolau Bach
Awel y Dol
Dull
Gwar Cwmsidan
ERLLYN
LLYN
DWYFN
PEN CLAW
MANERS
COR
DOL

could not begin a second trawl until the third had completed its first.

It may be suggested that one of the factors that contributed to the relative decline of the old system of pool allocation in the sixties and seventies was the arrival of the railway, which brought the Billingsgate fish market within very easy reach. An increasing number of fishermen owned a net and the coracle men forgot time-hallowed codes and fished where they would. The railway that came to Carmarthen in 1852, Whitland in 1854 and Cardigan in 1880, extended the fisherman's market and gave a powerful incentive to the netsmen of the Teifi to increase their activity and catches of salmon now that they would be transported with much greater ease to a mass market. Consequently, the local Teifi-side custom of salting and smoking salmon for preservation declined greatly.

Today the licence fee for a coracle net is £15.37½ and there has been a rapid decline in the use of the coracle as a fishing craft on the river. In 1861, a witness to the Commissioners of Salmon Fisheries described the number of coracles as 'legion', while another estimated that there were well over 300 coracles on the river. Another witness thought that there were between 200 and 300 coracles at Cilgerran alone, and that 'almost everybody in Cilgerran has a coracle'. At Cenarth, however, there were only 16 or 18 nets, while Aber-cuch had 10. The close season, by general agreement of the fishermen themselves, extended from 3 November to 3 March. Even in 1861 coracle fishermen had aroused the wrath of the rod and line anglers, for 'these coracles fish down the stream . . . and render the river unfit for rod or fly fishing'.[56]

After the eighteen-sixties, with the enactment of a number of Acts designed to limit and control salmon fishing,[57] close seasons and licensing of coracle fishermen were introduced. The close season varied, but it usually extended from 31 August to 1 February with a close time at weekends. Nevertheless, coracle fishing flourished throughout the last quarter of the nineteenth century and the first thirty years of the present century, with the result that in 1932 there were 24 coracle net licences, involving the use of 48 coracles at Cenarth alone, with a further estimated 52 netting permits below Llechryd Bridge.[58] By 1935, 33 netting licences had been issued.[59] The enactment of a bye-law, confirmed by the Ministry of Agriculture and Fisheries on 14 February 1935,[60] put severe restrictions on coracle fishing in the non-tidal section of the river. Today coracle nets may only be used by a person who immediately before the coming into operation of this bye-law was permitted to use a coracle net in the water above the bridge.[61] As a result of the 1935 bye-law, coracle fishing has virtually disappeared in the non-tidal section of the river. In 1952 16 coracle licences were issued for fishermen between Llechryd and Cenarth; in 1957 the number of licences had declined to 4, in 1970 to 1 and in 1971 no licences at all were issued.

The 1935 bye-law meant that coracle licences were not transferable as they are in the tidal reaches of the river, where at Cilgerran 5 licensees are allowed to fish from coracles. It has always been customary on the Teifi, for each licensee to be able to endorse two other names on his licence. Each endorsee (known locally as *gwas*) was expected to pay a shilling to the licensee for endorsement, and each one could operate the coracle without the licensee, as long as the licensee did not require the coracle himself. At Cilgerran, early in the twentieth century it was customary for an endorsee to apply for a licence of his own after three seasons as a servant. Undoubtedly, it is the rights of the rod and line fishermen that contributed indirectly to the disappearance of the coracle from the non-tidal reaches of the Teifi. As a report on salmon fisheries said,[62] 'There is no doubt that interest in angling has vastly increased of recent years and that the increase is continuing; anglers contribute greatly to the tourist industry and to the fishing tackle trade; some pay rates on sporting property to the local authorities in whose areas their waters lie and most contribute substantially to the funds of river boards through licence duties. On the other hand commercial fishing . . . employs comparatively few people and none of them is employed in the business all the time for there is a close season of at least five months; most of the fishing is done in public waters and there is therefore no contribution to local rates; and the amount of licence duty paid by netsmen is only in a few areas a noticeable contribution to the income of the river board.'

Today the coracle fishing season extends from 1 March to 31 August; night fishing is not allowed by the law of 1912, and the river is closed from 6am on Saturday to midday on the following Monday. The net is limited to a length of 20 feet and a depth of 3 feet 9 inches with a 2 inch mesh for the lint and 5½ inches for the armouring. Both coracle and net have to carry a licence number.

The salmon catches are today sold locally, but during the first quarter of the present century, Cilgerran had two fish merchants and Cardigan one, who paid regular visits to the river bank to purchase the salmon catches. In the nineteen hundreds, for example, merchants used to visit the river banks daily, paying about 2s 6d or 3s 0d a pound for the salmon.

Today all coracle fishermen are part-timers but until 1939 coracle fishermen working full days on the river were commonplace. During the closed season, many used to migrate to the coal mines of South Wales or were employed locally as quarrymen or rabbit catchers.

Coracles used on the River Nevern in north Pembrokeshire were similar to those of the Teifi, but even in the heyday of coracle fishing in the eighteen-sixties no more than four coracles, based on Newport, operated above the tideway. They and the seine netsmen of the estuary were at constant loggerheads for 'the men that fish the sand and those who fish the river are quite different men; the coracles belong to the men who fish the river and they are the men who give the most trouble; the men at the mouth of the river are fair fishermen'.[63]

THE TYWI CORACLE

On the River Tywi in the tidal reaches below Carmarthen town, twelve pairs of coracles are licensed to fish.[64] Most of the fishing is done at night between the old railway station at Carmarthen and the mouth of the river near Ferryside.[65]

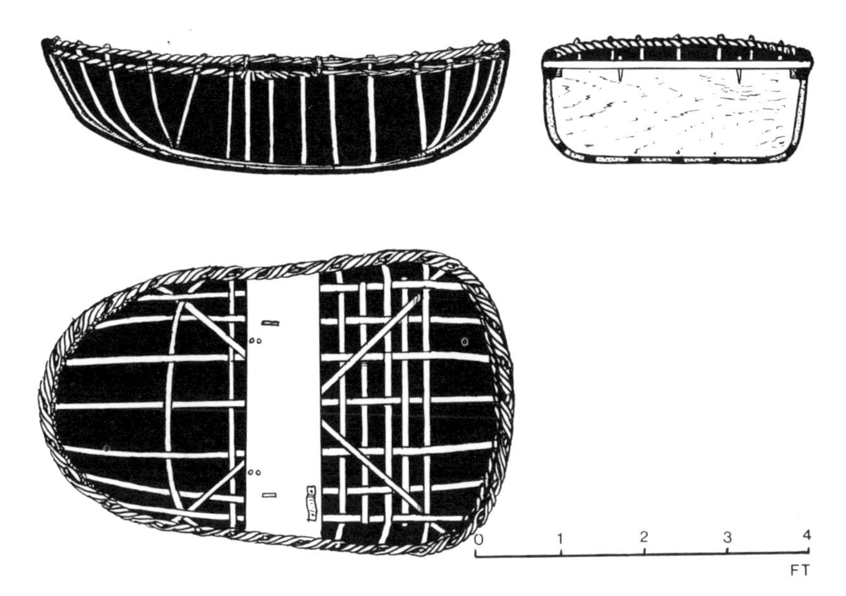

A Tywi coracle

The decline in coracle fishing on the Tywi has not been as spectacular as on the Teifi, for although in the early nineteen-twenties there was no limitation on the number of coracles operating on the river, the total number of coracles in use did not exceed 48 pairs. By the end of the decade, however, the number of licences had declined to 25[66] and by 1939 only 12 licences were issued. Since that date the number has remained constant, for fishery regulations state that new licences may be issued as long as the total number of licences does not exceed 12. Each licensee can endorse up to three other men on his licence, and any two of those whose names appear on the licence may fish at any one time. Nevertheless, compared with the total number of fishermen operating in the nineteenth century, the decline since 1918 has been substantial. According to one witness to the Royal Commission in 1861[67] so many salmon were caught by coracle fishermen that they were 'salted and sold as dried fish in the town and country and sent away. They used to be hawked about the streets' (of Carmarthen) 'by the fishermen's wives and by the fishermen and in the country along the side of the Towy'. Another witness to the Commission noted that the Tywi was being fished to such an extent that there was a scarcity of salmon in the Carmarthen district. This he attributed to 'the opening of the railways . . . Those who catch the fish now have a certain instead of uncertain market . . . Our local supply had been more affected by railways than any other cause'. It was estimated that in the eighteen-sixties there were about 'two hundred coracle men on the river who had no other employment but salmon and sewin fishing'.

The history of coracle fishing on the Tywi as on the other rivers of west Wales is obscure, but by the turn of the eighteenth century they were so common in Carmarthen 'that they cannot easily escape the observation of the tourist'.[68] Wyndham in 1781[69] describes how 'The fishermen in this part of Carmarthenshire use a singular sort of boat called coracles. They are generally 5½ feet long and 4 broad, their bottom is a little rounded and their shape nearly oval. These boats are ribbed with light laths or split twigs, in the manner of basket work, and are covered with raw hide or strong canvas, pitched in such a mode as to prevent them leaking. A seat crosses just above the centre, towards the broader end. They seldom weigh more than between 20 or 30 pounds. The men paddle them with one hand while they fish with the other'. John Evans on his visit to the Tywi[70] describes the coracle fishermen as 'the Caermarthen victuallers . . . on account of the quantities of salmon they supplied to the town'.

According to the *Second Annual Report of the Inspectors of Salmon Fisheries*[71] 'within the bay and tidal portions of the river no less than 400 men have for some years supported themselves and their families, two thirds of the year, on the produce of the salmon and sewin fisheries'. These included '12 to 15 long nets' as well as coracle nets. 'The coracles work in pairs', the Report continues, 'two of these hide boats with a man in each and a net stretched between them, drop down with the current. When a fish strikes they paddle rapidly together, and the fish and net are instantly landed in one coracle or the other'. The Tywi coracle men were described in the Report as being 'lawless and often aggressive, he poaches private waters for years and claims a prescriptive right; he uses violence if he be very strong, he threatens if his opponent be not so much weaker than himself as to make violence safe. A man of some property in this part of the country told me that they had encroached upon his water and taken possession of it; that he did not dare to interfere, for they would burn his stacks. This year, on the Towey, they went in a body to Ferryside, where the long-net men mostly live, attacked the men and destroyed their nets . . . The evil is at present grave'. Earlier, according to the *Reports on Education in Wales* of 1847,[72] some 1,500 people living in Carmarthen derived their living from the fisheries on the Tywi, 'and as their work was seasonal, they were frequently destitute on that account as well as because of their improvidence'.[73]

The Tywi coracle is longer and more elegant than the short, squat Teifi type. The following are examples of the dimensions of three typical examples:

Coracle A now at the Welsh Folk Museum (Accession Number 32-250/1) dates from the nineteen-twenties.

Coracle B and Coracle C were measured by Hornell.[74]

	A	B	C
Overall length	67in	68½in	64in
Maximum width (near far end)	43¾in	40½in	40in
Width (at seat)	40in	39in	39in
Depth (amidships to gunwale)	15¼in	15½in	16½in
Height from ground Front	15½in	16in	18in
Height from ground Back	13in	18½in	20in
Width of seat	10½in	10¾in	11in
Weight	32lb	31½lb	28lb

Tywi coracles differ in construction from the Teifi in the following respects:

1. Sawn ash laths rather than cleft willow rods are used for the framework of the coracle. As on the Teifi seven longitudinal

and seven transverse laths are used, but on the Tywi coracle 'the fore compartment is strengthened, not by doubling the three transverse frames, but by interlacing four short lengths of lath alternately with the first four transverse frames. The accessory laths do not extend up the sides'.[75]

2. A leather carrying strap *(strapen gwddwg)* replaces the twisted withy, hazel or oak *gwden* of the Teifi coracle.

3. There is no plaited band to reinforce the after end of the bottom framing as on the Teifi coracle.

4. Although in recently made Tywi coracles waterproof cloth coverings and a gunwale built up of sawn laths have replaced plaited hazel or willow, in older examples a visible plaited gunwale is an outstanding feature of the Tywi coracle. Although this gunwale is no deeper than that of the Teifi coracle, it has the appearance of being so, in that it is not covered with pitched canvas. On the Teifi coracle the calico is bound over the top of the gunwale with wire, but on the Tywi vessel it is turned back and tied with string to the lath frames below the top of the gunwale.

5. The Tywi coracle is equipped with a half round band of ash *(gwragen)* on the outside at the bow of the vessel to give protection to the front, where wear and damage is most likely to occur.

6. The after region of the seat is supported upon a solid wooden bulkhead *(astell orles)* held in place by three long pegs driven through the seat. The bulkhead is used for carrying the net or a salmon and differs considerably from the vertical railed seat support of the Teifi coracle. The bottom of the bulkhead is wired or tied to the laths.

To make a Tywi coracle, ash laths are preferred to withy laths, and from an ash pole 6 inches in diameter it is possible to obtain 30 laths. The ash is felled in the winter months and after splitting with a billhook or sawing in the local saw mills, the laths are placed in running water for two or three days, to facilitate bending. The seven longitudinal laths, the seven transverse laths and the two diagonal laths, each $1\frac{1}{8}$ inches to 1¼ inches wide, are interlaced in position on a flat plot of ground. Wooden boards and flooring are never used by Tywi coracle builders. 'Instead of fixing their position by superimposed weights' (as on the Teifi) 'the main crossings, those near where the laths are to be bent up are kept immovable by means of forked pegs, called hooks, driven into the ground in such a way that the forked ends straddle the crossings of the laths at all important points. From ten to sixteen of these hooks are used as available.'[76] The two central cross-laths are first of all bent up and the seat, usually of red pine or some other softwood, is inserted. The gunwale is then plaited with two plaits above and below the seat. The remainder of the ash laths are then bent and inserted in the gunwale. To make the gunwale, hazel rods 8 feet long and as 'thick as a finger' are required and about 45 rods of peeled hazel are required in order to make the gunwale of one coracle. These are placed in water for three or four days before plaiting. The wattled gunwale of a Tywi coracle consists of three distinct lines of wattling, but the final, wattled plait is not inserted until the coracle maker is perfectly satisfied with the shape of the framing.

Five yards of unbleached calico is required to cover the frame *(ofel)* of a Tywi coracle. This is cut to shape, usually a woman's task, and the calico is sewn onto the frame, ensuring that the cover is taut. On a sunny day, the coracle is tarred and a leather strap for carrying is inserted in the seat.

Unlike the Teifi coracle, the Tywi is equipped with a shallow, turned wooden baler, required for work in a broad tidal estuary where the water can be rough and broken. The Tywi coracle paddle is exceptionally long 'measuring overall 6 feet 1½ inches. The blade, including the well-sloped shoulder, is 2 feet 6 inches long, the loom 3 feet 7½ inches. The sides of the blade are parallel; in section it is slightly bi-convex. The loom is cylindrical in section and of equal diameter throughout; it is without either crutch or knob at the fore end'.[77] Paddles are made of either oak or ash, oak being preferred. The wooden club or *cnocer* up to 15 inches long is usually made of pitch pine, which many Carmarthen netsmen prefer to box-wood which is considerably harder and likely to spoil the appearance of a salmon. The *cnocer* and baler are carried under leather bands on the seat. A Carmarthen

A Tywi coracle fisherman, 1973

coracle is expected to last for at least four fishing seasons, but regular pitching is essential. Sometimes the upper plait of the woven gunwale has to be replaced, perhaps after two years, a process known as *ail godi* (re-raising). At the end of its useful life, it is customary to burn the coracle on the river bank.

On the Tywi it is customary for fishermen to supply their children with miniature coracles. 'When I was nine or ten years of age', said one informant,[78] 'I was allowed out on the river in my own coracle, a rope was tied around me and the seat of my coracle and this was attached to my father's coracle. I was taught how to paddle with a stroke similar to a figure eight. As I grew, I was given a larger coracle every year until I was fourteen or fifteen years of age, when I was given a full-sized coracle.'

Fishing on the Tywi is carried out at night, the beginning of a fishing session at twilight being described as *clyfwchwr.* As soon as 'seven stars appear in the sky' fishing can begin. The area fished consists of a large number of pools from the pump house near the old Carmarthen railway station to near the estuary. A trawl always begins nearer the right-hand bank of the river, described as *Ochor Tir* (the land side) rather than near the other bank of the river. At the end of a trawl the fishermen walk back to the starting point on the *ochor tir* side of the river, due to the large number of rivulets or 'pills' that enter the Tywi on the left bank. As on the Teifi, the river is divided into a number of casts with a series of well-established fishing pools between Carmarthen Bridge *(Bont Gaer)* and the sea. The principal ones are:

Pump House	Cerrig . . . Gwarddwyn[79]	Tywi Cottage
Brig y Bloiant	Brig y Brickyard	Tywi side. Pwll y bont
Gwar y bont	Gwar Meinen[80]	Ring y Myneni
Llyn Fach	Llyn Owen	Llyn Hywel
Banc y bont	Bwlch	Gwar ucha
Llyn Gou	Bwtri . . . Dwmde	Pwll Jini Ban
Cornel	Bwtshyn y bont	Gwar gwter
Gwar garw	Pibwr . . . Cwcwll	Gwar ucha pwll du
Pwll du	Gwar isha pwll du	Brig y gwter

Gwely Tomos	Y Gwter	Gwar gegen. Pwll y rifles
Gwter y gegen	Gwar Tom	Gwar Llangain
Gwar bach y gored[81]	Allt y Waddon	Gwar Shoni Bach: Banc yr Alma[82]
Gwar Glasne	Gwar gleiog	Gwter Shon Alban (now by-passed by river)
Pil rhoth[83]	Gwar Enoc	Ferry Point

Not all the casts that were used in the past are used today and the starting points are usually limited to Llyn Gou, Llyn Hywel Warfinen and Llyn Fach. The rigid rules of privilege and precedence, so apparent amongst the coracle men of Cenarth, Aber-cuch, Llechryd and Cilgerran, were never as well developed at Carmarthen. The first pair of coracles on the river bank with a net in place inside one of them is the first to begin the trawl. It is important that the net be inside the coracle or the pair will forfeit their turn. After the first pair has drifted with the ebb tide 'to meet the fish' that swim upstream, for about 200 yards, or in line with a pre-determined landmark on the river bank such as a certain tree or rock, one of the coracle men beats his knocker against the side of his craft. This is a signal for the second pair of coracles in the queue on the river bank to enter the water. In the past, it was considered wrong for the coracle men to talk when they drifted, but as soon as a fish hit the net, a shout of *'na fe na fe'* ('there he is, there he is') would come from one of the fishermen, and the pair of coracles would be hauled rapidly towards one another. The salmon is knocked on the head and placed in the box behind the seat. Before 1939 it was considered unlucky to hide a salmon on the river bank while the coracle men continued to fish, and it was not unusual for a coracle to carry as much as 80 pounds of salmon while the fishing continued. Today, however, a caught salmon is usually placed on the river bank for collection later as the fishermen walk back to the beginning of a trawl.

In hauling in a net, the partner nearest to the net is responsible for hauling in the net into his coracle and moving rapidly towards his partner. The left-hand partner usually, but not invariably, holds the reeving line and stapling line in his right hand and many fishermen have preferences as to whether they occupy the left- or right-hand positions. One fisherman, for example, says, 'I've always fished with my left hand and I am completely deaf in my right, for I can never feel a fish with it. If my partner is also used to fishing with the left hand, I have to drift downstream with my back towards my partner, so that I can hold the net in my left hand.'[84]

A fishing session on the Tywi extends from dusk to dawn, the session usually ending around 4am. The Tywi is regarded as a good river for sewin. In 1969, for example, the coracle men landed 1,470 sewin compared with 257 salmon.[85] So important is the sewin on the Tywi that specific names are used to denote the weight of fish caught. A sewin weighing 3lb to 20lb is described as a *gwencyn*; a *twlpyn* weighs from 2lb to 3lb, while a sewin of less than 2lb is described as a *shinglin*.

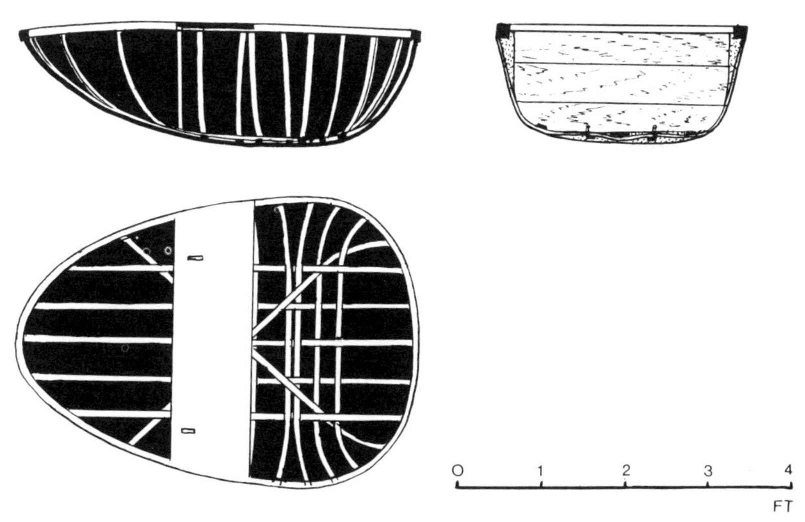

A modern Taf coracle

THE TAF CORACLE

The Taf is a short river, 32 miles long, that runs into the sea at Carmarthen Bay. Two pairs of coracles are licensed to fish in the lower tidal portion of the river 'between an imaginary line drawn straight across the said river from Whanley Point to Ginst Point and the main road bridge spanning the said river situate one mile or thereabouts below St Clears'.[85] The coracle men are all part-time fishermen based on the village of Lower St Clears. Although the fishermen are allowed to fish as far as the estuary of the river near Laugharne, most of the fishing is done between the two bridges at St Clears. In an enquiry, where it was proposed to change the boundary of coracle fishing in 1971, one witness complained of 'the operation of coracle nets in the River Taf, between the two bridges. The river there is very narrow and when the coracles are fishing not much space is left for fish to move upstream. The area for coracle fishing extends a long way downstream of the lower bridge and I do not know of any reason why the coracles could not fish in that stretch. During the last few years coracle netsmen have fished very regularly and have had good catches. Their use of motor cars as transport from one bridge to another has meant that they have been able to carry out a large number of sweeps daily.'[86]

It was maintained at this enquiry that the bye-law relating to the limits of coracle fishing was ambiguous for the river authority believed that 'the upper limit is meant to be the A4066 bridge approximately one mile directly south of St Clears' but the coracle netsmen 'have claimed that the upper limit is the A477 bridge approximately one mile south-west of St Clears'. As a result of this ambiguity, the Taf fishermen fished mainly in a stretch of river approximately one mile long between the two road bridges: a stretch of river considered totally unsuitable for the operation of coracles by the authorities. 'This stretch is at the upper limit of the tide and the channel of this small river is only 20 to 25 feet wide in places. Thus, when the net is fishing, few fish can move upstream because the net is across a substantial portion of the river. In addition particularly during low flows, sea trout stay in this stretch of river for long periods when a large proportion are caught by the coracle nets.'[87]

In evidence submitted by the coracle fishermen it was stated:

'There are two bridges in St Clears, both crossing the river Taf, and it has at all times been accepted by the Licensing Authority and the Netsmen, that the said description applies to the higher bridge, that is to say the bridge directly below and a mile or thereabouts from St Clears on the A477.

'If a ruling was given that the bridge so defined in the existing bye-law was the second bridge crossing the River Taf on the A4066 which has never been accepted by anyone concerned, it would in practice be tantamount to total prohibition of net fishing on the River Taf. This is evidenced by the fact that the river can be properly approached on its stretch below the higher bridge by public right of way, but river access would be prohibited if the lower bridge was defined as the commencement of the fishing length as it is bounded on both banks by privatley owned land which could prohibit passage by trespass. To realistically appreciate the position, it must be understood that the only fishable stretch of the present length is that part of the river lying between the two bridges. Below the second bridge the river bed is obstructed with tree stumps, weed growth, deep slime, and the adjoining marshes interlaced with deep wide drainage gulleys making coracle drift netting down stream and the consequent walk back an impossibility and beyond this, the length nears the river bar. It is clear, therefore, that to define the second bridge as the starting point of the permitted length is nothing more or less than total net prohibition and in that sense would be a ruling in excess of the powers of purely limitation order. When the limitation order restricting net fishing was introduced in 1931, it received the full co-operation of netsmen and their endorsees, and thus in a measure these very ancient fishing skills were preserved. It would therefore be a breach of faith on the part of the fishery constitution to so vary a bye-law as in effect to totally prohibit coracle net fishing in this area, which has already accepted the preservation of this ancient rural craft by a limitation order which insures practical continuance of coracle net fishing by the yearly issue of two licences only.'

A Taf coracle fisherman at Lower St Clears, Carmarthenshire, in 1937

The number of coracles on the Taf has remained constant since 1935, two pairs being licensed at that time. In 1933 three coracle nets were licensed, but in the eighteen-sixties 'between 40 and 50' were employed in salmon fishing on the Taf.[88] Earlier in the century a traveller to South Wales, Donovan[89] saw at St Clears 'a number of the poor inhabitants of the neighbouring cottages eagerly pursuing their customary occupations in the coracle fishery'. Donovan, although he had seen many coracles at Carmarthen, was particularly impressed by 'those feats of dexterity which are required in the management of such a capricious vessel' which characterized the coracle men of the Taf, or the 'Corran' as he called the river.

Although St Clears is no more than eight miles from Carmarthen, and although the Taf has a common estuary with the Tywi, the Taf coracles have suffered considerable modifications to suit the configuration of the river. Although the Taf coracle is similar in general shape to the Tywi, it is considerably heavier than the Carmarthen coracle. The front is flatter and the back more pointed and the wattled gunwale of the Tywi is replaced by planking. When Hornell carried out his survey in the nineteen-thirties he described the Taf coracle as follows.[90] 'The lattice part of the framework consists of seven longitudinal frames interlaced with either five or six transverse ones, all made of rough laths 1¼ and 1½ inches wide. No diagonal laths are present, but two or three short accessory laths to strengthen the bottom under the feet may be intercalated with several of the foremost transverse frames. The end of all the frames, bent up in the usual manner, are inserted, after being whittled down to cylindrical points about $\frac{3}{8}$ inch in diameter into vertical holes made at intervals in a broad gunwale frame of thin board; this takes the place of the wattled gunwale of the Teifi and Towy coracles . . . The seat is set flush with this gunwale, cleats below joining it to the gunwale frame, which does not extend beneath the seat. The partition supporting the after border of the seat is made of a number of broad strips of thin board set vertically at short intervals apart. These are nailed below to a basal bar extending across the bottom, while above they are nailed to a long cleat screwed to the underside of the seat. The cover is of calico coated with a mixture made by boiling 1lb of pitch with 1½lb of Stockholm tar. A round drainage hole is cut in the cover at the tail end, high up for the easy emptying of water when the coracle is taken out of the river.'

More recent examples of Taf coracles resemble the Tywi type although measurements and proportions are slightly different. An example made in 1970 and now at the Welsh Folk Museum consists of seven longitudinal and seven cross laths, the latter being all in front of the seat. Behind the seat to the slightly pointed back of the coracle there are no cross laths. Unlike pre-war coracles, the present day ones are equipped with two diagonal braces that cross in front of the seat. The gunwale, like the body laths, is made up of two layers of ash, thickly pitched, 1½ inches deep with the tips of all the laths inserted between the two layers. Attached to the deal seat are a strap for carrying and another for holding the knocker in place. Three planks at right angles to the seat act as a carrying box, while the seat itself is merely nailed to the top of the gunwale and flush with it. In use, the fisherman has to sit well forward of the centre of the coracle to provide equilibrium, for on the ground the Taf coracle has the appearance of being particularly lop-sided with the back almost touching the ground and the front rising at a sharp angle. The paddle is of the Tywi type, 5 feet long with a blade 2 feet long with parallel sides and gently sloping shoulders. The knocker is of crabapple wood or box wood, and although a strap is provided for carrying it on the seat, by tradition Taf coracle fishermen preferred to carry the club in their pockets.

It seems that considerable modification in the design of the Taf coracle took place in the late nineteenth century, so that recent examples closely resemble those of the Tywi. One informant said[91] that to build a Taf coracle of the older variety a naturally curving branch of a tree, usually an apple tree, was cut and split in half to form the fore-part of the gunwale. The two sections were fixed together with a cleat iron. Another branch was treated in the same way to form the rear gunwale. Unlike the Tywi coracle, it was the gunwale of the Taf coracle that was formed first and this was bored with a series of holes for receiving the laths of the coracle frame. Seven longitudinal and seven cross laths were prepared by cleaving willow with a froe. These were pointed, laid down on the floor and the pointed ends inserted in the hole in the gunwale. A seat was tacked on top of the gunwale and the coracle was covered and tarred in the usual manner.

The following are examples of the dimensions of two Taf coracles:

Coracle A was made by Edgeworth Evans of Lower St Clears in 1970, now at the Welsh Folk Museum (Accession Number F71-70).

Coracle B was measured by Hornell in 1937[92]

	A	B
Overall length	59in	57in
Maximum width	39¾in	42in
Width at seat	39in	39in
Width at back	38in	37in
Depth (amidships)	18½in	Not given
Height (front)	15½in	Not given
Height (seat)	19in	12½in
Height (back)	15½in	Not given
Width of seat	11in	Not given
Weight	31lb	33lb

CORACLE FISHING IN OTHER RIVERS

The use of the coracle as a fishing craft has declined very rapidly in recent years and today coracle fishing is limited to three rivers only, the Teifi, Tywi and Taf in west Wales. In the nineteen-twenties and thirties, coracles were to be found on many other rivers such as the Dee, the Eastern Cleddau in Pembrokeshire and the Severn, while in the late nineteenth century other rivers such as the Wye, Usk, Conway, Dyfi, Nevern and Loughor had coracle fishermen. Each river had its own specific type of craft, but many examples have disappeared without a record being made of them.[93]

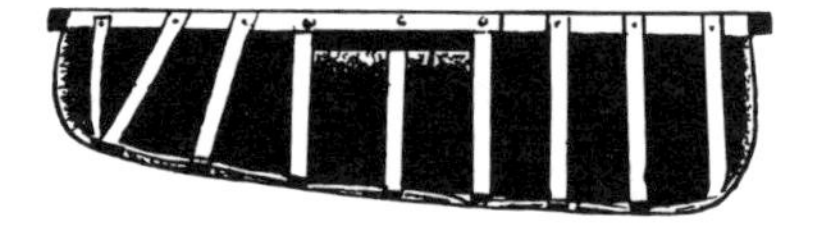

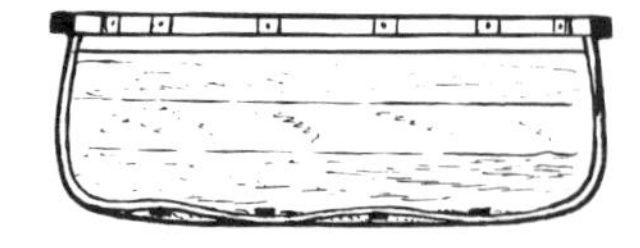

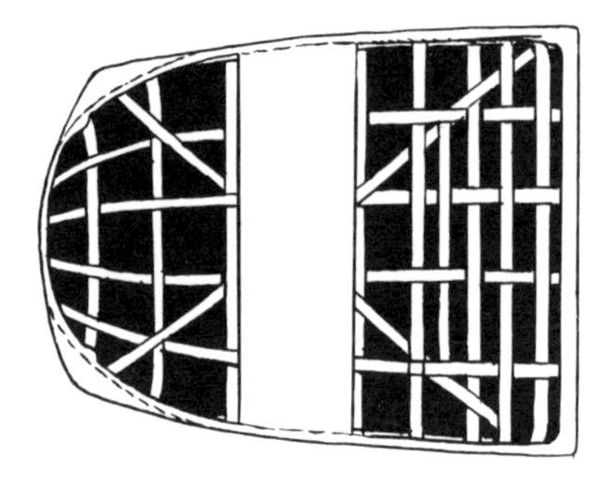

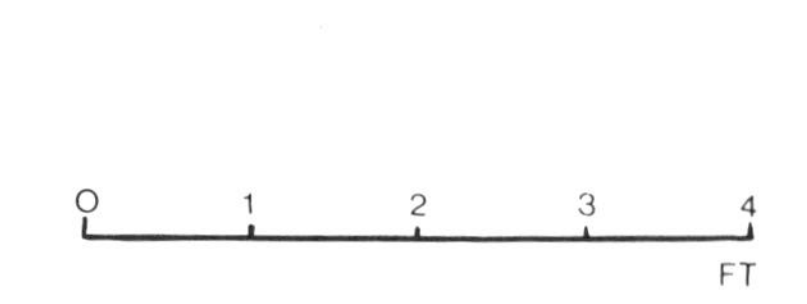

A Cleddau coracle

Cleddau

Coracle fishing persisted on the eastern branch of the Cleddau until 1939, when a single pair of coracles was licensed to fish. Even in the eighteen-sixties, the number of coracles on the Cleddau never exceeded 6 pairs above Llawhaden bridge, and 6 to 10 below.[94] The coracles were used frequently for poaching. By 1930, the number had declined to three and by 1934 one pair only was found on the river. Coracle fishing ceased completely in the early nineteen-forties.

The Cleddau coracle was closely related in type to that used on the Teifi and had 'the same short squat form; the same deep, wide, square fore end and short rounded "tail" as that of the Teifi. Clear evidence of its origin is afforded by the form and size of the paddle. This is 4 feet 3 inches long, made up of a short blade, 2 feet by 3¼ inches and a loom ending in a transverse claw grip, identical with that of the Teifi paddle'.[95]

Although no Cleddau coracle has been preserved, it is possible

A Cleddau coracle man at Blackpool, Pembrokeshire, in 1937

to deduce details of its construction from photographs.[96] The framework of the coracle consists of six laths, made of sawn ash 1½ inches wide running the length of the coracle and nailed to the inner side of the planked gunwale. At right angles to these laths are inter-woven three cross laths, six in front of the seat to provide support for the fisherman's feet and two behind the seat. All the laths are bent upward and nailed to the gunwale. The gunwale itself consists of a lath, about 1½ inches wide, bent into a U-shape to form the sides and back of the coracle frame. Nailed to this is a piece of wood, approximately 1½ inches square, that forms the bow of the coracle. The front framing of the Cleddau coracle therefore is far more squared than that of the Teifi. Two diagonal laths that intersect beneath the seat completes the frame. The seat, with leather carrying strap and leather knocker strap, is placed midway along the length of the coracle. Two pieces of wood from about 14 inches to 18 inches long and about 1½ inches square are bolted on either side of the seat to the gunwale. The deal seat in turn is bolted to the under side of these two pieces of wood. A solid bulkhead is screwed to the seat about 3 inches from the back end and this runs to the floor of the coracle being supported by the crossed diagonal laths of the framing. Hornell said that the Cleddau coracle was covered with Hessian canvas, coated with a mixture of pitch and tar. The measurements he gives are as follows:

Length	– 52 inches	Breadth	– 40 inches
Depth (front)	– 14 inches	Depth (amidships)	– 13 inches
Depth (back)	– 10 inches	Width of seat	– 12 inches

Usk and Wye

A coracle was last used for angling, in the Usk, around 1930, but undoubtedly coracle fishing was well known on the river and its tributaries as well as on Llangors Lake *(Llyn Safaddan)* until the late nineteenth century. In Camden's *Brittania*[97] of 1586 'Llyn Savaddan . . . In English 'tis called Brecknockmore . . . well stored with otters and also Perches, Tenches and Eels, which the Fishermen take in their coracles'.

It seems that although nets were used on the Wye and Usk for

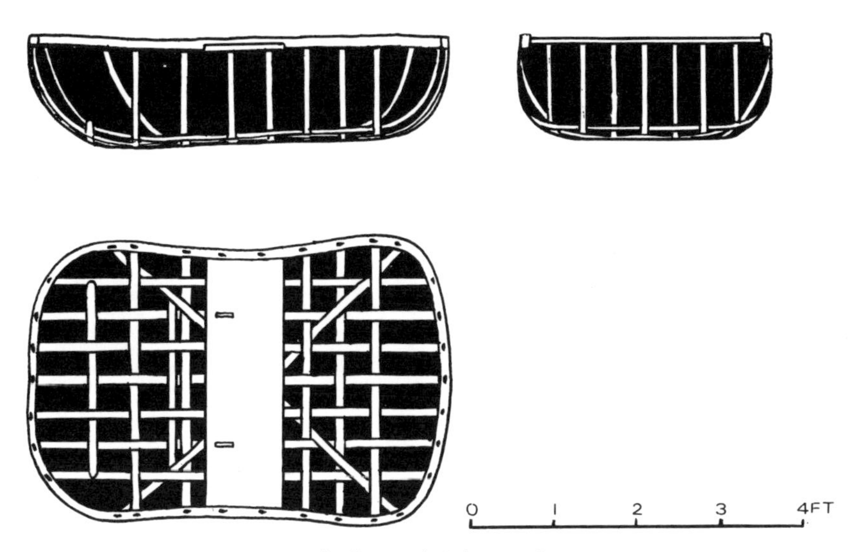

A Wye and Usk coracle

salmon fishing, the coracle was more widely used in angling. Sir John Hawkins in his 1760 edition of Izaak Walton's *Compleat Angler* for example[98] notes that 'the men of Monmouthshire made use of "a thing called a Thorrocle or Truckle" when fishing, in this case fly fishing for grayling, which were not easy to get at without a boat or wading. In some places it is called a "Coble", from the Latin "Corbula", a little basket; it is a basket shaped like the half of a walnut shell, but shallower in proportion and covered on the outside with a horse's hide; it has a bench in the middle and will just hold one person, and is so light that the countrymen will hang it on their heads like a hood, and so travel with a small paddle which serves for a stick till they come to a river; and then they launch it and step in: there is great difficulty into getting into one of those Truckles: for the instant you touch it with your foot, it flies from you, and when you are in, the least inclination of the body oversets it.'

A mid-nineteenth-century writer[99] noted that on the Usk

Coracles on the Monnow, circa 1900

'many of the inhabitants gained their livelihood a great portion of the year by netting and angling. During the season, near the town of Usk, ten or a dozen fishermen were to be seen carrying their coracles on their backs in going to and returning from their avocation . . . Their shape resembled the section of a walnut shell, the length was about five feet and the breadth about four, with a seat placed across the centre; they were made of thin hoops crossed, with very strong basket work edges, and covered with strong coarse canvas thickly coated with pitch . . . the fisherman might often have been observed to work his paddle with one hand while he conducted the net with the other, at the same time holding a line in his teeth . . . On the banks of the Usk, Wye and other fresh water rivers, these coracles were to be seen hanging at the doors of many of the cottages'. After 16 July 1866 it became illegal to fish for salmon without a licence on the Usk[100] and each coracle licensee had to pay an annual fee of £2 to use a net in a fishing season that extended from 1 March to 31 August.

On the Wye and Monnow coracle fishing was also widely practised until about 1914, but after that date the number of coracles in use declined rapidly, and there is no evidence to suggest that coracles persisted after World War I.[101] In the Ross area they were widely used in the nineteenth century. A traveller in 1799,[102] for example, noted that 'During the course of the navigation from Ross, we passed several small fishing craft called Truckles or Coricles, ribbed with laths or basket-work, and covered with pitched canvas.'

Down river at Monmouth they were equally well known. 'Many salmon are caught at this place which is five miles [upstream] from Monmouth' said one writer in 1805.[103] 'Here we saw several boats, called coricles, peculiar to this part of the river . . . we saw two men going out in their coricles to fish. Each man lays hold of one end of a net, about 20 yards long, and paddles down the river till they feel a strike. They then haul it up as quick as possible and draw it on shore. They paddle along at a great rate, and put us much in mind of what we read concerning the Indians in their canoes.'

The fact that the Wye coracle fishermen used an armoured net is confirmed by a witness to the Royal Commission on Salmon Fisheries in 1861.[104] The mesh of net was 2½ inches and in addition to drifting with the flow of the stream, as in west Wales, Wye fishermen used coracles for the process of 'bushing'. A bush, a hole or rock in the river, was surrounded by a net and the salmon driven out of its hiding place by poking with a pole. For the more orthodox technique of drawing with a 'truckle net' a witness to the Commission said: 'There are two coracles apart from one another and about 16 yards of net. One man has a running line; each man holds one end of the net with one hand, and paddles with the other to keep the coracles as far apart as the net will allow: directly they feel a fish, the man at one end lets go and the other draws up and the fish is bagged and drawn into the coracle.'

As on the Usk, the use of the coracle for angling persisted

longer than its use for netting on the Wye. One Ross angler 'used a reel, but others only had a large cork bung on a short piece of cord attached to the rod butt, and on hooking a fish, the lot was heaved overboard. The rod, etc, played the fish, and directly he rested the fisherman paddled after his bung and gave it a pull to start the quarry off again'.[105] Hornell describes the rod and line used for coracle fishing in the Hereford district.[106] 'The rod is of elm, short and stiff, and shaped like a billiard cue, tapering upward from a stout and heavy butt. Its length is barely 8 feet. The line is of horsehair, about 24 feet long, tapering from a diameter of 3mm down to six hairs at the end.'

The coracles used on the Usk and Wye were identical in construction and were usually constructed by the fishermen themselves. The Monmouthshire 'truckle', as it was called locally, showed remarkable resemblance to those used on the Tywi, which may suggest a common origin. It differs, however, in that the section behind the seat is longer than the section in front of the seat and the seat is not set near the centre as on Tywi coracles. The gunwale of the Monmouthshire coracle had none of the sheer displayed by the Tywi. An example of a Wye coracle preserved at the Hereford City Museum[107] and made by William Dew of Kerne Hill, Ross on Wye, about 1910, has a broad and deep fore-end and a rounded stem with the laths curving up gently at the back. The sides are parallel and the gunwale almost horizontal. The dimensions are as follows:

Length	– 60 inches	Height (fore end)	– 14½ inches
Width (fore end)	–39½ inches	Height (at seat)	– 14½ inches
Width (at seat)	– 38 inches	Height (stern)	– 15½ inches
Width (back)	– 39 inches	Weight	– 28lb
Width of seat	– 11 inches		

The coracle has seven longitudinal ash laths and six transverse laths, inserted in a woven hazel gunwale. Between the sixth and seventh cross lath and behind the seventh are two short accessory laths to give added strength in the long tail of the Wye coracle. A pair of diagonal laths cross just in front of the seat, giving added support to the solid bulkhead under the seat. A leather carrying strap is fitted.

The paddle of a Wye and Usk coracle is similar to that of a Tywi, being 5 feet long and with a parallel-sided blade from 18 inches to 21 inches long. The loom is straight and cylindrical.

Hornell describes the method of building coracles adopted by Mr A. C. Morgan of Monmouth.[108]

'The framework consisted of seven longitudinal and seven, or rarely eight transverse laths crossing one another at right angles, with two diagonals, all arranged after the Towy fashion. When a coracle was to be begun Morgan states: "I used to go to the sawyer and say 'Rip me out a set of laths'." These had to be split from willow logs [sally-wood]; sawn laths were not considered satisfactory, width 1¾ inches to 2 inches. These laths were soaked in water for two days before use. When judged pliable enough, they were laid on the ground and interlaced at the proper distances apart. Then the main crossing points were secured either by means of forked pegs driven into the ground [Carmarthen method] or were held down by weights [Cenarth method]. This done the ends of the laths were bent up and secured in position by plaiting withy bands around them at gunwale level. The ends of two of the transverse laths a little in front of mid-length were the first to be bent up; these were passed through slots in the ends of the seat. Amidships rigidity was obtained by inserting a solid deal partition of template form beneath the seat. This was tied below at two points to the lath framework and above at two corresponding points to the seat, the sewing passing through two small holes forward of each of the slots made for the carrying strap.

'The withy gunwale plaits were arranged in such a way that if a plait was begun on the left behind the seat and circled forward clockwise, the end of its half-circle had to finish on the right in front of the seat; the complemental semi-circle reversed this procedure – it started in front of the seat on the right and then circled round the stern to end and interlock with the beginning of the fore-end plait, behind the seat on the left.'

A wooden mallet for killing the fish was carried, made preferably of apple or pear wood. A bailer was considered unnecessary.

The cover was of stout calico and was usually called the 'Hide' of the coracle. The fishermen's wives made the cover; after stretching it over the frame and lashing it on below the upper plait of the gunwale, a coating of a mixture of pitch and coal tar was applied on the outside. Afterwards the coracle was taken (in Monmouth) to the Gas Works and left in one of the retort houses for 24 hours in order that the mixture should thoroughly permeate the fabric of the cover. After being brought home, a second coating of the same mixture was applied. The mixture was carefully tested; a stick was dipped into it and brought out with a small blob of the stuff on its end. After a few seconds it was passed between the fingers and then, if it stretched and did not crack, it was considered to be of the right consistency.

Dee

In 1920 coracle netting rights on the Dee were bought out by the Dee Fishery Board, who since 1903 had made a conscious attempt to limit coracle netting by not issuing new licences on the

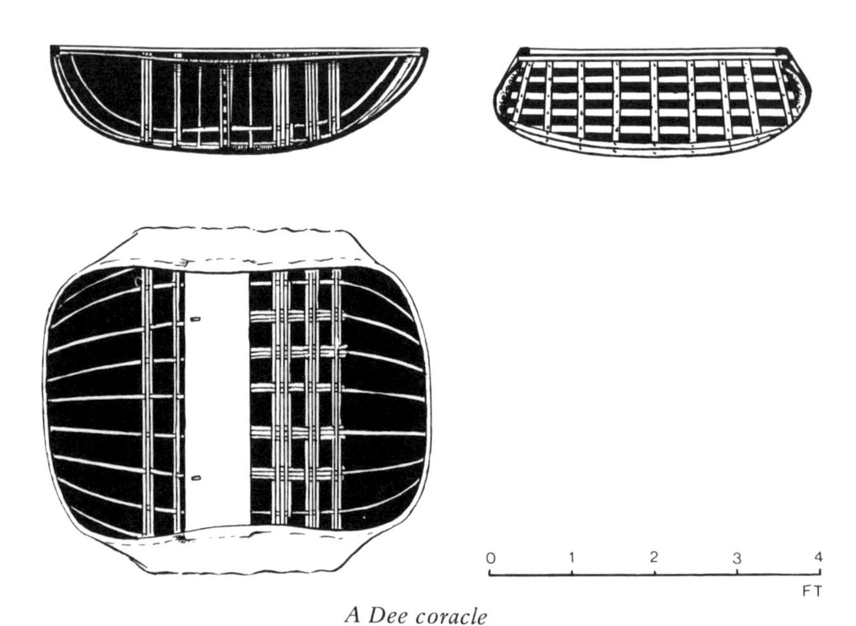

A Dee coracle

death or retirement of a licensee. In 1920 the three remaining nets operating on the river in the Bangor-on-Dee—Overton district were bought out for £1,000, and as a result coracle netting on the river ceased. Nevertheless coracles remained in use for angling until the nineteen-fifties. 'A coracle is almost a necessity on the rough, rocky, middle reaches of the Dee', says one writer,[109] 'when owing to the force of the current and deep, hidden ledges, and clefts in the rocky bottom, wading is impossible in many places and no other type of craft, not even a birch bark canoe could possibly be used.'

In the nineteenth century, coracle netting was undoubtedly widely practised on the Dee, especially in the upper reaches of the river. One witness to the Commissioners on Salmon Fisheries in 1861, for example,[110] said 'It happened frequently that several hundreds of the men would go out very early in the morning with coracles on their backs, pass over the mountains and come some distance down the river, taking all they could catch with very fine nets. They all met at a certain public house, where large white baskets were filled with what they had obtained and sent to the Liverpool and Manchester markets.'

Even in the eighteen-sixties, coracles were a considerable source of worry to the authorities, for the 15 pairs that operated between Bangor and Erbistock weir were 'fished upon suffrance for many years'.[111] In 1862 says another report[112] 'The coracle fishing has been rather increased this year, and this mode of fishing somewhat endangers the future of the river . . . Unless some restriction be placed upon this class of fisherman, it is evident that they will increase as the fish increase . . . Complaints have been frequently made to us that the coracle fishing has discouraged the proprietors from taking steps to improve the river, because they would derive no benefit thereby, and increase what they already consider to be a nuisance'. During the last three decades of the nineteenth century, coracle netting was certainly discouraged on the Dee and whereas 27 licences for netting were taken out in 1871, by 1884 the number had decreased to 15 licences and by 1895 to 12. In addition, in 1888 the Fishery Board limited the length of the salmon fishing season by putting back the opening of the season from 1 February to 31 March, 'with the view of making certain that Kelts had a fair chance of returning to the sea'.

Although the coracle persisted on the Dee until fairly recently amongst anglers, it was hardly ever used in the nineteenth century for that purpose 'except for catching eels with a "bob" – a bunch of worms upon worsted'.[113] In the nineteen-forties and fifties many of the coracles used by anglers, especially in the Llangollen district, were made to accommodate two persons and the vessels were wider than many of those used on the other rivers. Usually they were about 55 inches wide by 57 inches long. Oddly enough two-man coracles persisted longer and were more common on the Dee than the one-man variety, for as Hughes-Parry says: 'Most coracles are made to accommodate two persons, although a certain number of single coracles are still in use.'[114] These double coracles weighed about 40lb and with them 'it is possible to shoot rapids and dodge in between out-jutting ledges in the fastest and wildest stream, holding on by gaff and paddle to some outcropping ledge or rock; and one can tuck oneself in in perfect safety with a rush of wild white water on each side. By using a coracle one can therefore fish places that could never be reached either by wading or throwing the longest line from the bank'.

There were two distinct types of Dee coracle.

1. The Lower Dee coracle used mainly in the Bangor-Iscoed-Overton area.

2. The Upper Dee coracle used between Bala as far as beyond the mouth of Ceiriog. Most of these coracles were to be seen in the vicinity of Llangollen.

The Lower Dee coracle, an example of which is preserved at the Welsh Folk Museum, has sharp in-curving sides, especially below the seat, and a broad flattened bow with a more rounded stern. The framework is formed of interlaced ash laths, that are much thinner and narrower than those used in any other coracle type. There are nine of these longitudinal laths, each no more than ¾ inch wide, and these laths are strengthened by a number of supplementary laths that are inserted, one on each side of the five central splints, in front of the seat where the coracle fisherman's feet will rest. The transverse laths too are greatly increased in number and the majority are arranged in compound sets, each set corresponding to one of the broad transverse laths of South Wales coracles. The first of the transverse laths from the bow of the vessel consists of three separate splints, the second, four and the third, five. The gunwale is formed of two ash laths bent to shape with the deal seat inserted between the two, while the *cratch*, the box-like space under the seat, is formed by a stout pillar under the seat and lattice work. This cuts the coracle into two separate sections – a fore section 30 inches long and an after compartment 24 inches long.

When not in use the coracle is set upright upon its 'tail'. To prevent deformation of the slender framework when thus upended, a stout median wooden bar is inserted between the seat and the centre of the stern gunwale.

A paddle, a carrying strap of the usual type, and a thong loop passed around the central pillar below the seat are the only accessories. The paddle is 5 feet long and of unusual elegance. The blade and loom are of equal length; the blade tapers evenly from the broad distal end, and were it not for the merest trace of shouldering, would merge insensibly into the cylindrical loom. No crutch is present; in its place we find the end of the loom encircled by an iron band, notched or slotted at one side. When the owner slings the coracle on his back, he passes the end of the loom through the thong loop passed around the seat pillar; the notch serves to prevent the loop from slipping off.

The principal dimensions of a Lower Dee coracle from Overton are:

Length	– 55 inches	Length of seat	– 38 inches
Maximum width (front)	– 38 inches	Width of seat	– 11 inches
Maximum width (centre)	– 35 inches	Depth (at seat)	– 14 inches
Maximum width (back)	– 47 inches		

Hornell describes the method of construction as follows:[115]

'The two sets of laths which are to form the framework are arranged as usual on the ground or preferably on a wooden flooring where they are kept in relative position after interlacing by weights or by tacking down.

'The ends of the transverse laths on one side are then bent up and locked between two stiff rods running longitudinally, each lath being tied to these two embracing rods. The lath ends on the opposite side are similarly treated; this done, the ends of the two opposed pairs of rods are connected by cords at the distance apart which is to be the eventual width of the coracle at gunwale level. The ends of the longitudinal laths are similarly treated and held in place by cords running fore and aft. The result is that the framework appears of the form of a rectangular basketwork trough with the four sides not joined together at the corners.

'To facilitate bending, the laths are sometimes thinned slightly at the bends.

'When the curved laths are considered to be sufficiently set, the latticework seat-partition is placed in position and wired at several points along the bottom edge to the laths below. This done the two lower gunwale hoops are nailed in position, one outer and the other inner to the laths and about an inch and a half below what will be the eventual gunwale edge. The ends of the laths embraced by the paired rods are now released and the seat may be put in, its ends passing over and beyond the lower gunwale hooping. The ends of three of the transverse laths are passed through slots in each end, a procedure which causes the waist-like appearance when the coracle is completed. The seat is further secured by being wired at intervals through paired holes to the upper edge of the partition below.

'The upper gunwale hoops are next added, one on each side of the projecting frame ends, which are now cut off flush. These upper hoops pass over the seat ends.

'The coracle is now ready to be covered with calico. The edges are reflected over the gunwale and tacked on. After receiving a coating of the usual pitch and tar mixture, an extra inner gunwale hoop is added to hide and protect the turned-in edge of the coracle.'

The Upper Dee coracle, used in the past from Bala Lake to below the mouth of the River Ceiriog, differs considerably from

Dee coracle at Llangollen, 1933

the type of coracle used in the Overton-Bangor districts. They were used until recently in the Llangollen district and most were designed to carry two people. For this reason the coracles were given a more accentuated bilge so that in appearance the Llangollen coracle was almost square, with the ends considerably sheered.

The Llangollen coracle differed from the Bangor-Overton type in that broad, planed laths were used for the frame construction. Nine of these were interlaced with sixteen transverse laths to form a stout framework that was further strengthened by the addition of four diagonal laths. Seven stout squared pillars inserted into cross-bars formed the cratch and support for the seat. Unlike the Lower Dee coracle, the Llangollen type did not possess the distinctly pinched waist, so characteristic of the coracle of Bangor and Overton. Methods of carrying also differed, for the Llangollen coracle was never equipped with a carrying strap. 'The customary way of carrying it is to support the seat across the shoulders, steadying the coracle with the hands gripping the sides, close to the fore-end . . . The method of carrying supported by the hands over the head is a comparative recent innovation. Old photographs show the coracle being carried in the orthodox manner by a leather strap across the breast and shoulders, with the paddle resting horizontally on the left shoulder, with its loom end inserted within the interior of the coracle.'[116] A short 4 foot paddle was used on this type of coracle. The blade 22 inches long was parallel sided, being 4½ inches wide connected to a cylindrical loom without a crutch.

The following are the dimensions of a typical Llangollen two-seater coracle:

Length	– 57 inches	Beam - Bilge	– 53 inches
Width of seat	– 11 inches	Beam - Front	– 33 inches
Length of seat	– 39½ inches	Height - front and back	
Depth to top of seat	– 14 inches		– 20 inches
Weight	– 50lb	Height - centre	– 17 inches

The method of construction was described by Hornell as follows:[117]

'After the two sets of laths have been interlaced in the usual manner on a plank flooring, a wooden roller about 3 feet long by about 6 inches diameter, having an iron pin running through it, is placed lengthwise over the laths on each side, and secured to the floor by iron brackets . . . The two rollers are arranged at a distance apart of about what the eventual gunwale beam is to be. Discarded rollers from an old mangle are suitable for this purpose.

'The projecting parts of the transverse laths, after a preliminary soaking with warm water, are bent up and tacked against the upper part of the rollers on the outer aspect, their ends sloping inwards to form the tumble-home type of side characteristic of this design. After being left for some time to set the bends, the projecting ends, at the proper level, are nailed between a pair of lower gunwale hoops, each composed of two half-hoops, in the Bangor manner, but instead of being arranged horizontally, the half-hoops are fixed with such a sheer towards the ends of the framework that their ends cross one another obliquely amidships. The two rollers are now removed and the seat with its two sets of pillar supports are put in and secured in place; this permits of the nailing on of the upper circumferential hoops to form the gunwale proper. A cover of sailcloth is stretched over the frame; an overlap at each corner strengthens what are the weakest places.'

Conway

James Hornell, when he carried out his survey of coracles in the nineteen-thirties, was of the opinion that coracles were unknown on the Conway and all those in use at the time had 'been imported, two at least from Llechryd on the Teifi'.[118] Hornell was incorrect in his assumption, for coracles were widely used on the river, at least until 1914, and those coracles were quite different from those used on other rivers in north Wales. In 1887, for example, John Jones of Tanrallt, Betws-y-coed, who was licensed to use a basket trap on the Lledr[119] which resulted in litigation in that year, was also licensed to use a coracle for salmon fishing on the Lledr as well as the Conway. His coracle is now at the Welsh Folk Museum, St Fagans.

Undoubtedly coracles were used in large numbers on the Conway, at least from the sixteenth century,[120] and Michael Faraday on his visit to the area in 1819 said: 'Here and there on the river we saw fishermen in their coracles: little vessels something like a washing tub squeezed by a door into an oval form; a board is put across the middle in which two men sit, one each way and whilst one paddles the other casts the net.'[121] It seems, therefore, that two-man coracles were used on the Conway as they were on parts of the neighbouring Dee. By about 1840 the use of coracles as fishing craft on the Conway had declined very greatly. In evidence to the Commissioners on Salmon Fisheries in 1861[122] a witness said that coracle fishing had ceased in the Llanrwst area, and added that 'below Llanrwst, where the tide comes . . . they were only used where the men could not land their nets; they never used them below Trefriew . . . they fished about seven or eight miles above the tideway and a little lower down for sparlings'. Coracle fishing during the last quarter of the nineteenth century and the first decade of the twentieth seems to have been limited to the Betws-y-coed district, well up river from the tidal reaches.

The Conway coracle preserved at the Welsh Folk Museum[123] is unusual in that the framework consists of broad, 3 inch strips of cleft ash interwoven to form a solid-looking frame. The longitudinal laths are 6 in number and these are crossed by 9 other laths. The gunwale is also of ash, each lath being 2 inches wide, and in between the two oval hoops forming the gunwale the ends of the long and cross laths are inserted and nailed in place. On the outside of the gunwale a hoop, similar to that used by coopers for cask making, is nailed to the timber. Above the fifth cross-lath a heavy balk of timber is nailed and two pillars for supporting the deal seat are inserted in this. The space below the seat does not form a carrying box for fish as in some coracles, but the pillars were used by the fishermen for carrying the coracles on their heads; the pillars acting as hand grips. The Conway, like the Welshpool coracle, is equipped with a carrying strap and the paddle is a long piece of ash without

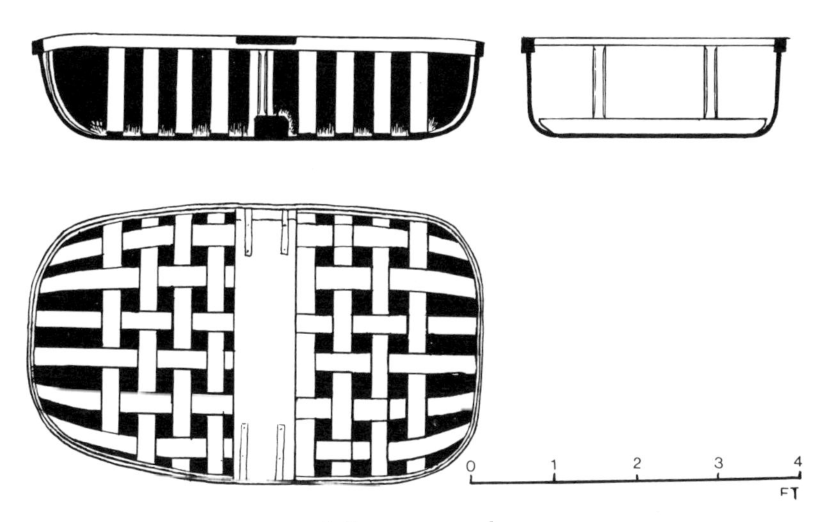

. *A Conway coracle*

shoulders to the blade and without a hand grip at the top. The blade itself, flat at the front and rounded at the back, is 22 inches long and gradually merges with the round tapering loom of the paddle. The deal seat of the coracle, supported on two wooden pillars, is also strengthened by two pairs of metal straps, nailed to its top and inserted between the two ash laths of the gunwale.

The principal dimensions of the Conway coracle are:

<table>
<tr><td>Length</td><td>– 66 inches</td><td>Height from ground (seat)
– 14 inches</td></tr>
<tr><td>Maximum width
(at seat)</td><td>– 40 inches</td><td>Height from ground (front)
– 15 inches</td></tr>
<tr><td>Width (front)</td><td>– 38 inches</td><td></td></tr>
<tr><td>Width (stern)</td><td>– 38 inches</td><td>Height from ground (rear)
– 13¾ inches</td></tr>
<tr><td>Depth (at seat)</td><td>– 14 inches</td><td></td></tr>
<tr><td>Weight</td><td>– 35lb</td><td></td></tr>
</table>

Severn

A River Board bye-law of 1890 severely restricted and the Salmon and Freshwater Fisheries Act of 1923 spelt the doom of coracle netting on the Severn. Undoubtedly before 1923 more coracles were used on the Severn than any other river in Britain, for coracles were used for a distance of approximately sixty miles along the river between Welshpool in Montgomeryshire and Bewdley in Worcestershire. They persisted until 1939 on a stretch of river extending to about thirty miles between Shrewsbury and Arley, but of course, not officially for salmon netting.[124] Today, a single part-time coracle man, Eustace Rogers, at Ironbridge in Shropshire follows a tradition that is said to have been in his family for over three hundred years.

Unlike the coracles of West Wales, the Severn coracle was never a vessel used specifically for salmon netting and was widely used 'for ferrying, angling, laying lines and the carriage of stone and brick sinkers required for the lines and of the large wicker traps employed in eel fishing'.[125] Since bridges on the Shropshire section of the Severn are few and far between, the coracle was widely used by the inhabitants of Severnside as an easy means of crossing the river. An 'Ironbridge' coracle for this reason is considerably broader and more manoeuvrable than any other type. 'As many as four people were ferried across the river on one occasion by an old man . . . The passengers stood around the paddler clutching his shoulders and each other.'[126] So considerable was the demand for coracles in the Ironbridge district at the beginning of the present century that a school was established where the principles of coracle manipulation were taught.

When coracles were used on the Severn there were three distinct types: 1. The Ironbridge coracle, 2. The Shrewsbury coracle and 3. The Welshpool coracle.

The Ironbridge coracle as made by the late Harry Rogers until his death in the early nineteen-sixties is an almost oval, bowl-shaped craft, approximately 57 inches long and 36 inches wide. The gunwale is almost horizontal, being 14 inches above the ground and the seat is usually 9 inches wide. The coracle weighs no more than 27lb and the framework consists of ten longitudinal and nine cross laths or 'splints', each 1½ inches wide and made of sawn ash. Short lengths of diagonal splints are inserted at the four extremities of the frame to give added strength to

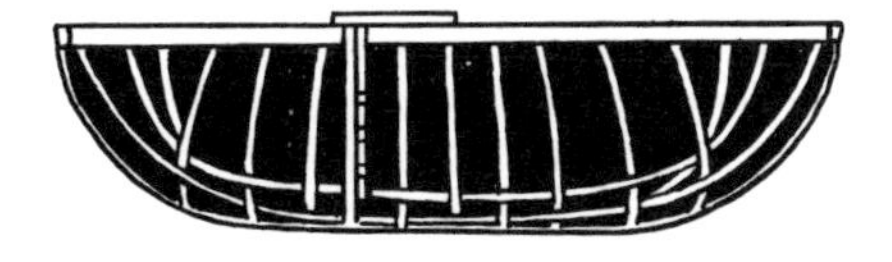

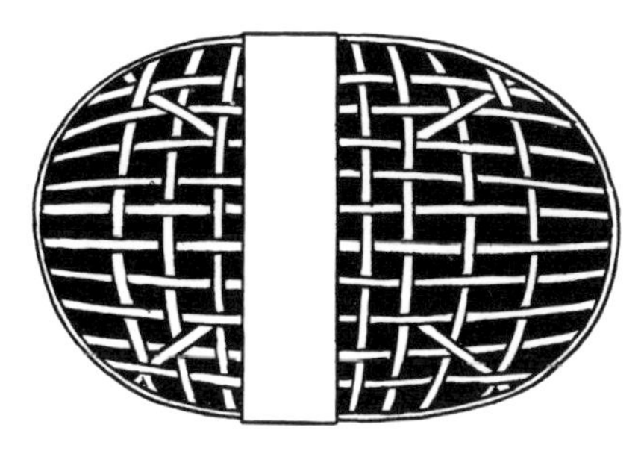
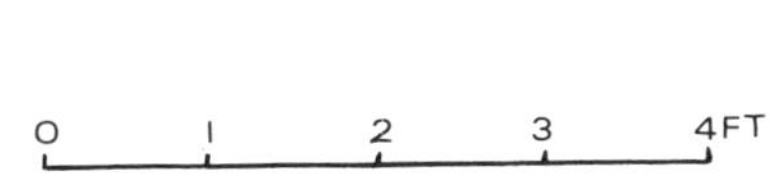

A Severn (Ironbridge) coracle

the frame, but these four diagonals are only 20 inches long and do not run the length of the frame. The gunwale consists of two half-hoops, overlapped and spliced together at the end to form an oval.

Hornell describes the method of constructing the Ironbridge coracle as follows:[127]

'The laths or "splints" are bought ready sawn to the proper thickness; they average 8 feet in length. Those for the frames are a fraction under ¼ inch thick, while those for the gunwale hoops are slightly over this thickness. Before use the laths are soaked with hot water to make them supple. When ready, those that are to run fore-and-aft are laid upon a wooden flooring or some sort of wooden platform such as an old door and spaced apart at regular intervals; this done, the transverse laths are interlaced and then, to keep them in place, the laths at the four corners are tacked down to the flooring or platform.

'Prior to this an oval hoop formed of two half-hoops, overlapped and spliced together at the ends, has been prepared of the size and form to be taken by the gunwale. The ends of all the laths, hitherto lying prone on the flooring, are now bent upwards and tacked but not clenched to the outer side of the oval hoop, at a height of about 14 inches from the ground. After this, strings are passed in various directions across the hoop and between the upstanding ends of the bent-up laths, in order to bring them to the proper curvature. These strings prevent the laths from springing out of curve but do not hinder some of them from being pulled inwards, so it becomes necessary to run "stays" outwards from their ends to the plank floor to obviate this. These outer stays are particularly required at the corners, which are the most difficult parts of the frame to shape correctly. In this condition and under constant adjustment of the controlling strings, the framework is left for several days for the bends to become set in position. At the end of this time, an outer and permanent hooping – the so-called "skeleton hoop" – is put on; the first or temporary one being removed thereafter. The projecting ends of the ribs are next cut off level with the top edge of the skeleton hoop. This done, the frame is set free from the tacks holding it to the flooring and turned bottom up in order that the laths may be tarred on their outer side. The framework is also ready to be covered with its "hide" of unbleached calico. As bought, this is 1 yard wide, so two widths are overlapped 3 or 4 inches and sewn together. This seam runs down the centre line of the bottom. When adjusted in position the free margins are turned over the edge of the skeleton hoop and tacked on at short intervals: any excess is trimmed away.

'At this stage the bulkhead, which is to furnish the median support of the seat, is put in. This done, pitch and tar, roughly in the proportion of 1 quart of tar to 2lb of pitch, are boiled together and a coating of the mixture applied over the outside of the cover.

'The following day the two remaining gunwale hoops are added, one on the inner side of the rib ends, the other on the outer side of the skeleton hoop but separated from it by the fabric of the cover. To do this four half-hoops are made by bending laths to the shape required. A cord adjusted between the two ends of each half-hoop keeps them in shape till set. Before, however, fitting the inner hoop, two short strengthening bars are fitted at each corner of the frame, as these places are weak owing to the frame ends diverging here rather widely. The stern half of the inner hoop is put in place first, four iron screw-clamps or "dogs" being used to hold it in position while being nailed to the skeleton hoop by 1 inch paris points. In the same way one of the outer stern hoops is put on at the after-end of the frame. Finally the forward half-hoops are put on, one within the frame ends, the other outside the calico cover. Care is taken to allow sufficient overlap at the junction of each set of half-hoops.

'All that remains to be done is to fit the seat in position. This is laid athwart the coracle a little abaft the centre; its ends rest on the gunwale at each side, and each end is secured thereto by three angle ties of iron, 1½ inches long, each screwed in place, the outer ones with one screw through each arm, the middle one with two screws, as these screw into the bulkhead bar, whereas the others screw into the hoops of the gunwale. The seat is further secured by screws or by nails passing through it into the heads of the three bulkhead pillars. Two slots are cut in the seat, each about 8 inches from either end, for the carrying strap of leather, and a thin leather thong-loop is put around the centre seat prop; through this loop the head of the paddle is passed on the fore-side when the coracle has to be carried, thereby relieving the pressure upon the chest. As a final touch some owners paint the outer gunwale hoop and the inner faces of the lath frames.'

The paddle used at Ironbridge is almost spadelike in shape, with a broad blade 16 inches long and 7¼ inches wide with a cylindrical loom screwed to it. A straight hand grip 4 inches long at the top runs at right angles to the loom. The whole paddle measures 40 inches long. An older type of paddle in use in the nineteenth century was in one piece. 'The blade', says Hornell, 'is elegantly tapered from the broad distal end upwards to its junction with the loom, which is circular in section. The crutch or grip is a bluntly rounded expansion of the head of the loom. Length 5 feet; blade 30 inches long by 6 inches at the outer end; loom 3½ inches in circumference; crutch 3 inches long. Locally the loom is termed the "stale", while the crutch is the "casp".'

The following are the principal dimensions of a coracle made by Harry Rogers of Ironbridge in 1955 and now at the Museum of English Rural Life, University of Reading:

Length – 57 inches Height of gunwale – 14 inches
Width – 43 inches Width of seat – 11 inches

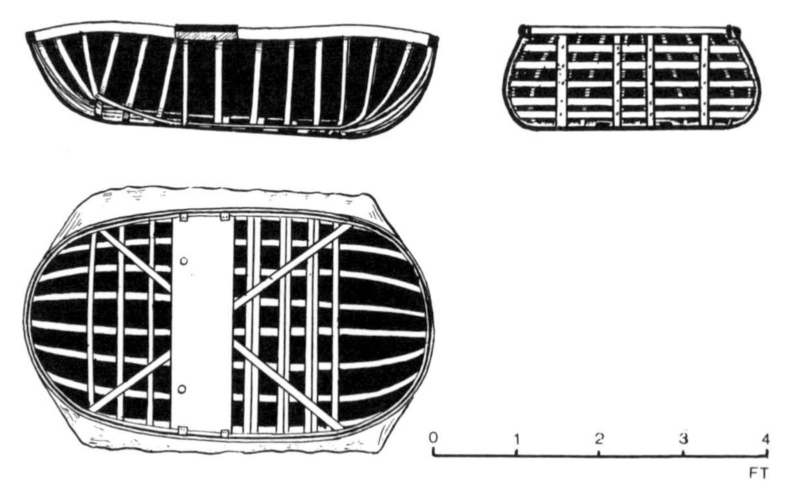

A lower Severn (Shrewsbury) coracle

Coracles were in use on the Severn at Shrewsbury until 1939 and they were mainly used at that time by anglers. Hornell points out that some of the coracles he saw in use in the nineteen-thirties had been made on an elaborate wooden mould around which the framework was built. Nevertheless although the mould was used by a Shrewsbury boat builder from about 1880, only six coracles were built on it, and the technique of construction usually adopted by the coracle makers of the town resembled those of Ironbridge.

The Shrewsbury coracle is less oval in shape than the

Ironbridge type and like the Wye and Tywi coracles it has a distinctly broad bow and a more pointed stern. The framework is noticeably in-curving towards the gunwale, and the front and rear compartments of the coracle are separated by a grating of four uprights crossed at right angles by five ash laths. These laths are of the same width and thickness as those used for the frame and gunwale of the coracle. The gunwale itself consists of three thicknesses of laths nailed together to form a sturdy frame. A half-round hazel hoop is tacked on to the lath gunwale all the way round to give added strength to the framing. Eight lengths of sawn ash splints, each 1½ inches wide and ¼ inch thick, form the longitudinal membranes of the frame. Above them but not interlaced with them are another eight laths at right angles to the longitudinal laths. When the laths cross, a nail is inserted and each lath in turn is nailed between the splints of the gunwale. A pair of diagonals crossing just in front of the seat run the whole length of the coracle, but these again merely rest on the framing and are tacked to them rather than being interlaced. Undoubtedly where sawn ash is used, there would be a danger in cracking the timber if the thin strips of timber were interlaced. The front compartment of the coracle is strengthened with three short cross-pieces tacked to the frame to give added strength at a place where the fisherman would rest his feet.

An example of a late nineteenth-century Shrewsbury coracle at the Welsh Folk Museum[128] has the following dimensions:

Length	– 58 inches	Width (at seat)	– 33 inches
Depth (at seat)	– 13¾ inches	Width of seat	– 8¾ inches
Height of gunwale (back)	– 10½ inches		
Height of gunwale (at seat)	– 15 inches		
Height of gunwale (front)	– 16 inches		
Weight	– 31lb		

In shape and design the coracles of the Welshpool area had close affinities to those of the Wye and Usk. Salmon netting on the upper Severn was prohibited as early as 1890 and coracles were used in the Welshpool area until their final disappearance in the nineteen-thirties for such tasks as retrieving ducks when shooting and for angling. Before 1934 coracles were also used for setting night lines[129] but as a result of legislation at that time, setting night lines, like netting salmon, became illegal on the Severn.[130]

Hornell describes the method of constructing a Welshpool coracle as follows:[131]

'The materials required to construct the framework consist of seven ash slats [laths] 7 feet in length and eight others of 5½ feet length; all must be "rent" or cleaved by hand with a hoop

A coracle on the Severn at Shrewsbury, circa 1930

shaver to a width of from 1⅛ to 1½ inches by ¼ inch thick; also four more carefully fashioned slats for the gunwale. These obtained, three of the longer ones are laid down upon an old door . . . the outer ones at 2 feet 7 inches apart toward one end and 2 feet 4 inches toward the other. Two short slats are laid transversely across these, 3 feet apart . . . and the points of intersection secured temporarily in position by being tacked through to the door beneath. The remaining four longitudinal slats are next laid down, and then the rest of the cross slats, six in number, are laced in and out of the seven longitudinal ones at about equal distances apart. This gives an open basketry with rectangular meshes about 6 inches along each side.

'Having the four corner points tacked down to the door flooring below, the ends of two of the median longitudinal slats, after softening with hot water, are bent upward and nailed to the outer side of an ovate gunwale frame. A couple of the cross-slats are similarly treated. With these guide frames in position the rest are easily worked into their respective places.

'The inner gunwale band referred to is made up of two lengths of wide ash slats about 1¾ inches wide, each bent into an oval form and joined to its fellow by an overlapped joint. The two oval bands differ markedly in radial curve.

'To stiffen each of the four corners of the frame, a short length of slat is placed diagonally between the gunwale and the second slat crossing from front and side; each is 23 inches long. Other strengthening pieces are four accessory "foot" battens, each 3 feet long, placed alternately with the first five transverse frames and nailed over them. When all the framing is in position, the seat is put on. This is a board 8 inches wide, 2 feet 10 inches long at the fore-side and 2 feet 8 inches along the afterside. Each end rests upon the edge of the gunwale band and is screwed to a short batten nailed against its inner face at the place where the ends of the two sections overlap. To hold the bottom stiff, the seat is supported at one-third its length from each end by a stout cylindrical rod about 16 inches in length; its lower end rests upon one of the bottom slats.

'At this stage the frame is set free from the holding down nails

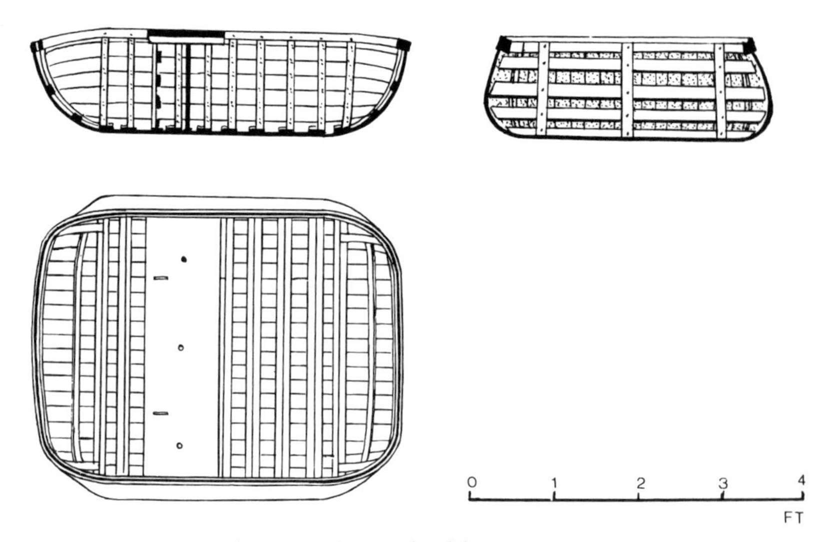

An upper Severn (Welshpool) coracle

and is turned bottom up, to have its cover of stout calico put on. This done, a coat of pitch and tar is applied both to the inner and outer surfaces. Last of all an outer gunwale slat band is nailed to the outer side of the frame ends which are thus enclosed between an outer and an inner gunwale band.'

The ash paddle of the Welshpool coracle has a long narrow blade without distinct shoulders and measures 51 inches long. One side of the blade is flat, the other convex and there is no hand grip at the top. The Welshpool coracle is not equipped with a carrying strap, and for transporting the vessel 'the coracle is lifted by both hands with the bottom upwards; thus inverted the flat of the seat is brought to rest on the left shoulder; then the paddle is placed across the right shoulder with the blade inserted under the seat in order to take part of the weight. The net, in the days when this was used, was carried on the top of the inverted bottom'.[132] It seems that the horizontal method of carrying coracles, without carrying straps being employed, was limited to the upper Severn above

Shrewsbury, the Conway and the upper reaches of the Dee. In all other districts, coracles were equipped with carrying straps of leather or withy.

The dimensions of a Welshpool coracle measured by Hornell were as follows:[133]

Overall length	– 58½ inches	Depth (front)	– 16 inches
Maximum width	– 36 inches	Depth (back)	– 13½ inches

A Leighton Bridge fisherman, who was 78 years of age when recorded in the early thirties by Stanley Davies, described the method of construction as follows:[134]

'This type of coracle has no sheer; if any sheer were given to the fore-end it would be difficult to pull in a salmon out of the net. It is 4 feet 9 inches long, by 3 feet 3 inches wide, and 18 inches deep at the seat. The pointed end is the stern and is called the back. The bow is called the fore-end. The centre of the seat is 2 feet away from the back. The user sits facing the fore-end. The frame is made of ash, riven by hand with a hoop shaver (a cooper's tool). The slats are 1½ inches by ¼ inch. The interwoven framework is made of 7 slats lengthways and 8 slats broadways, with a short slat added in each corner.

'The coracle is made as follows: An old door is placed on the ground. The slats are laid on and interwoven, and each nailed down in two places to the door. Only slats which have been riven can be interwoven. If the more modern method of using sawn laths is adopted the laths must be nailed together with copper nails. The ends of the slats are then softened with hot water, and bent up to meet the inside rim of the gunwale. The slats are then nailed to the gunwale. The nails should be flat-headed clog nails, as they are soft enough to be clinched, but they are difficult to obtain. The gunwale is of ash, and is 13 feet 4 inches long. Owing to the difficulty of obtaining such a long length it can be in two pieces which meet under the seat. In addition four slats are laid on the bottom of the coracle to take the pressure of the user's feet. There are also two round pieces of timber one inch in diameter one end of which is screwed to the underside of the seat and the other end to the bottom of the coracle to distribute the weight of the seat. The frame is then covered with calico and waterproofed inside and out with a mixture of 2lb of pitch and 1lb of tar. The outer rim of the gunwale is then nailed on. Two short strips are nailed to the gunwale to help to support the seat. The seat is 8 inches wide and is fixed last of all. Alongside the left hand side of the seat is kept the "priest", a stout oak stick about a foot long used for stunning the salmon. It is hung just below the gunwale in two loops of leather.

'The paddle 4 feet 3 inches long, has the edges of one side of the blade champhered, and this side must be kept "next to the water", which you are drawing towards you; otherwise you will find yourself out of the coracle and in the water. There are two ways of using the paddle. One is to pull the coracle through the water by placing the paddle in the water in front of you, and, using both hands, working the paddle in a figure eight, keeping the champhered side of the blade towards you, "next to the water" which you are drawing towards you. When netting you place the paddle in the water on the right hand side of the coracle, and tuck your arm around it, with the top of the shaft resting on your forearm, and your fingers over and equally divided each side of the shaft. The left hand is then free to handle the net line.

'I prefer to carry the coracle horizontally and inverted, lifting it up and letting the flat of the seat rest on the left shoulder; then I place the paddle on my right shoulder, and let the blade fit under the seat to take part of the weight. The net is thrown over the top [really the inverted bottom] of the coracle, which avoids the user getting wet from the net. If you carry the coracle on your back by means of a strap looped to the seat, the wind is liable to fill the coracle and blow you off your feet.

'Materials used to construct a Severn Coracle.

ASH:

7 Lengthway slats, 7 feet 0 inches	)	
8 Broadway slats, 5 feet 0 inches	)	all 1¼ inches x ¼ inch
4 Corner pieces, 2 feet 0 inches	)	
4 Pieces under feet, 2 feet 0 inches	)	

4 Pieces for gunwale, 7 feet 0 inches x 1¼ inches x ¼ inch
2 Seat struts, 1 foot 6 inches x 1¼ inches

DEAL:
Seat, 3 feet 6 inches x 8 inches x ¾ inch, planed, self edge
2 Seat supports, 8 inches x 1½ inches x ¾ inch

OAK:
"Priest", 12 inches x 2 inches diameter

ASH PADDLE:
4 feet 3 inches x 4½ inches x 1 inch

'To complete the coracle the following materials are required – calico, pitch, tar, leather loops, copper nails, iron clog nails.

'In transit a hole can be stopped by a clay clod, taken from the river bank and held in place with the foot. Large holes were repaired with patches of calico painted over with hot pitch. Small holes were repaired with hot pitch, but now it is more convenient to warm a stick of gutta-percha with a match and apply it to the hole. Constant painting and repairing steadily increases the weight of the coracle.'

Although the use of the coracle as a fishing craft in the present century has been limited to the above rivers, in the nineteenth century their use was far more widespread. In south Wales, for example, bowl-shaped nearly square coracles were used on the River Loughor at Pontardulais, a stream that became heavily polluted during the last quarter of the century.

The Loughor was a particularly good salmon and sewin river in the late eighteenth and early nineteenth century, says Donovan[135] in describing Swansea market. 'Half a dozen females seated upon the panniers of their ponies . . . rode hastily down the market place with a supply of sewen . . . conveyed from Pontardulais, about ten miles to the westward . . . abounding with fish during the summer, being caught in the coracle fisheries by peasantry.'

The Dyfi in mid-Wales, another river that became heavily polluted with effluence from lead mines, was also a coracle-

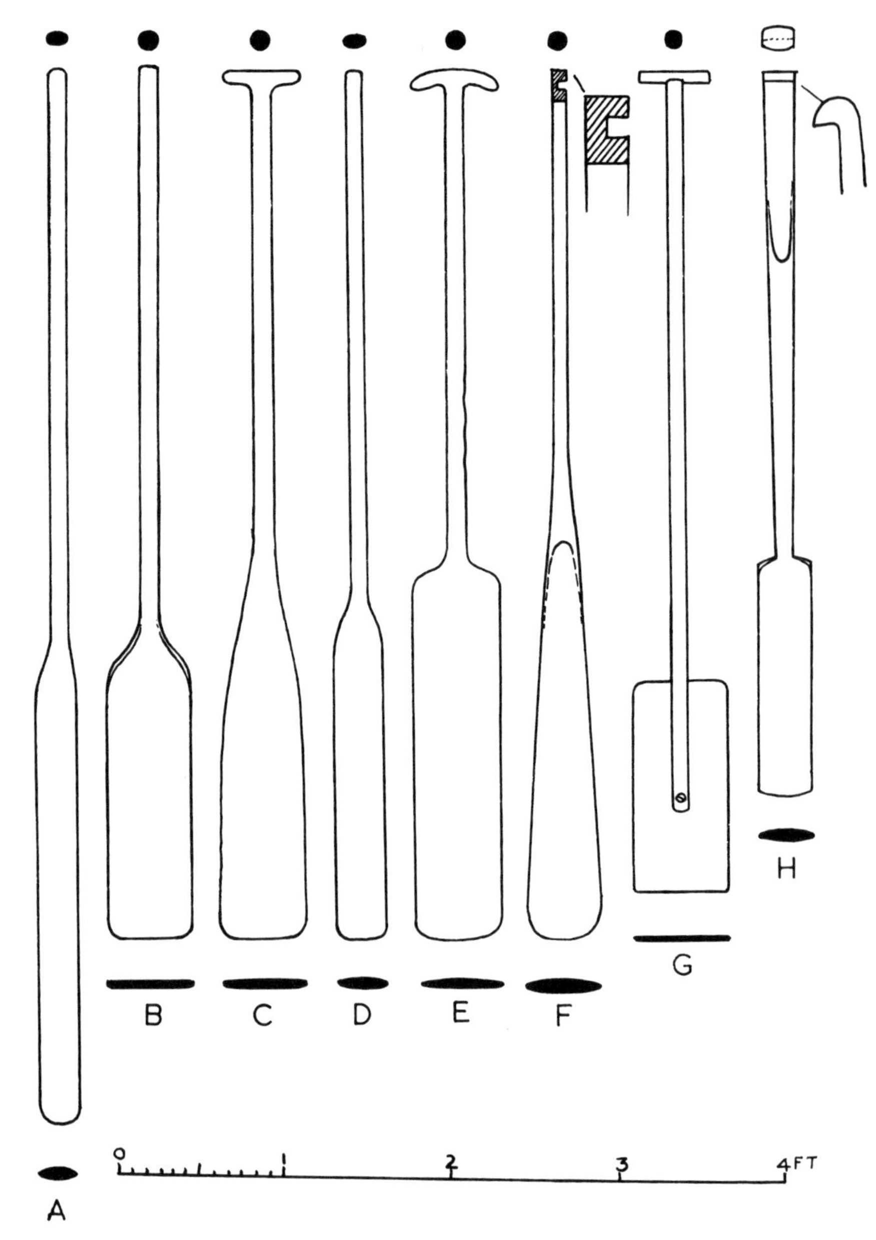

Types of coracle paddle: A Tywi; B Wye and Usk; C Severn (Ironbridge); D Taf; E Severn (Shrewsbury); F Dee (Bangor); G Severn (modern Ironbridge); H Teifi

fishing river. In 1800 Bingley described coracles that were from '5 feet to 6 feet long and 3 feet and 4 feet broad, of an oval shape, so light that one man may with ease carry them on his shoulder'.[136] By 1861 coracle netting, due to pollution and the influence of the Dovey Fishery Association, had ceased and the 24 coracles operating on the river were declared illegal, although a few years previously 'there was an immense deal of coracle fishing'.[137]

For notes to Chapter 7, see pages 314-19

8 Drift Netting

The term 'drift net' includes 'those gill nets which are not fixed to the bottom, but are attached to a boat or to a series of buoys, free to move with the wind or tide; the wall of nets tailing out up wind or up-tide, whichever has most influence'.[1] The vertical wall of the net is set, and any fish that strike this will become caught in the meshes, unless they are small enough to pass through, or too large to get their heads into the mesh. In drift netting, therefore, it is important that the net is drifted at right angles to the flow of the tide, so that any fish in the river enter the mesh directly and do not bounce off it, if the net is at an angle to the ebb or flow. In practice the fish either 'jams the mesh tightly round itself and is held firmly by it, or if the gill covers pass through the mesh, even a lightly jammed fish is trapped, as the gill covers prevent it backing out'.[2] Within the limits allowed by law, it is usual to use nets of different mesh according to the size of fish expected. 'A 40lb salmon would not get its head into a 1½ inch mesh, and so would not be caught; a 2lb sea trout would go through a 4 inch mesh and would escape.'[3] In the Wye, for example, the drift netsmen use a 3½ inch meshed net for grilse, while later in the year nets with a mesh of 5 inches for catching larger fish are employed. If the 2¼ inch mesh specified by the River Authority were employed throughout the season, most of the large-sized fish caught would escape, as they would not be able to get their heads into the mesh and would bounce off the net.

In drift netting, one end of a net is attached to a floating buoy or leaded staff while the other end remains fixed to the boat. The head rope is corked and the foot rope leaded to keep the net upright in the water, and one of the main tasks of the drift fisherman is not only to ensure this, but also to ensure that the net, 300 yards or more in length, is at right angles to the flow of the tide. Drifting with a strong ebb or flood tide demands considerable strength in the use of oars to make sure that the drifting boat and net are constantly in the correct position for fishing. If the net should drift more rapidly than the boat, tangling can occur, and the net is certainly not performing its duties efficiently. Although in all districts engines are used for taking the boat to the correct fishing area, the engine is stopped before the actual drifting commences. In fishing, both boat and net are at the mercy of the wind and tide.

In England and Wales generally, the stronghold of drift netting seems to be the north-east coast, in an area that comes under the jurisdiction of the Northumbrian River Authority. Much of the fishing is carried out along the coasts by teams of drift netsmen, each team consisting of not more than three members. The nets used are 600 yards long with a mesh of 1½ inches from knot to knot. The salmon fishing season extends from 26 March to 31 August. Many of the north-eastern fishermen combine drift netting with using fixed engines in the form of 'T-nets'. The T-net is in effect a double-anchored trap with a leader running from the foreshore. This was legalised in 1962 by a special parliamentary order and at that time most of the salmon commercially taken in north-east England were taken by the T-nets. With the discovery of mono-filament nylon drift nets, the T-net has been rapidly superseded, the reason being that synthetic nets enable drifting to take place in the daytime and in fine weather conditions, whereas drifting was only effective at night time and with rough sea conditions with hemp nets.

The number of licences and endorsees in 1971 numbered 324 but with two net limitations in force from 1 March 1972, the number of licences had been reduced to 170 in the 1972 fishing season; most of these are licensed to fish with T-nets and drift nets.[4] In 1966 25 licensees with 56 endorsees were allowed to fish with drift nets between the northern tip of Lindisfarne Island to Boulby Craggs in Yorkshire, but a further 33 persons with 84 licensees were allowed to use drift nets as well as T-nets. In 1970 a total of 154 licences with 389 endorsements were allowed in the north-east, divided as shown in Table 9[5] *(below).*

TABLE 9

AREA	TYPE OF FISHING PERMITTED	MEN ENGAGED Licensees	Endorsees
Whole Area From Goswick Sands in north to Souter Point in south	T-nets and drift nets from Howick Burn to St Mary's Island; drift nets only in remainder of area	56	142
South Area From Church Point Newbiggin to Souter Point	T-nets and drift nets from Newbiggin to St Mary's Island; drift nets only from St Mary's Island to Souter Point	8	11
North Area From Goswick Sands to Howick Burn Mouth	Drift nets only	6	9
Wear & Tees Area From Souter Point to Boultby Cliffs	Drift or hand nets	84	227
	Total	154	389
	Total number of men engaged in salmon and trout fisheries	543	

In north-west England drift nets, known locally as 'hang nets', are used in the area that comes under the jurisdiction of the Cumberland River Authority. Two licences are issued for the Solway, two for the Cumberland coast and one for the Esk estuary.

Each team of netsmen consists of two members and the nets, which have a minimum mesh of 1 inch and are attached to a leaded-end staff or pole. In use the boat and net are allowed to drift with the tide. Most of the nets used are 300 yards in length, but there is no real limitation on the length, although extra licence fees are payable for each additional 40 yards above the 300 yards.[6]

To the south of the Lake District drift netting, known locally as 'whammelling', is carried out on the Ribble and Lune, two rivers that come under the jurisdiction of the Lancashire River Authority. R. S. Johnson describes the method as follows:[7] 'Whammelling, or drift netting, is done from a boat. On the Ribble estuary, where the river is confined between training walls, the drift nets are limited to 150 yards in length; on the Lune, nets up to 320 yards in length are fished. The nets are about 9 feet deep, leaded along the bottom, corked along the top, so that they float vertically in the water. Varying meshes are used; on the Ribble the minimum is 3¼ inches, on the Lune 2½ inches, but above these minima the deciding factor is the size of fish the netsmen believe to be in the estuaries at any particular time. If they think there is a run of large summer fish they put on a 3¾ inch mesh net; if, on the Lune, a number of grilse show themselves in July, they may put on a 2½ inch mesh net.

'The net is piled in the bows of the boat. The boat starts from one side of the river. One of the two men in it rows across to the other bank while the other pays out the net as they go. The end of the net is marked with a leaded pole or a buoy. The aim is to keep the net as straight as possible so that it drifts down at its maximum length, gill-meshing the salmon as they run up. Changes in the speeds of the currents, cross eddies and varying depths soon put kinks in the net, and it is not often that you see a whammel on the Lune fishing its full length. The river varies in width, too, so that the net is constantly being hauled in or paid out, and with a strong ebb running, there is a deal of pulling to be done. On the Ribble, where the training walls canalise the estuary, the netsmen can, on a calm day, shoot the net, and, when it is fully paid out, cast off and row back to the end first shot to straighten it, but they cannot do that on the Lune.

'Drift nets are limited to six on the Ribble, and to twelve on the Lune. On both rivers there are bye-laws restricting the area in which the nets may be fished. On the Lune the area begins at Baithaven Buoy, and there the boats congregate as soon the tide has ebbed to start fishing. Conditions vary from day to day and tide to tide, but normally it is not worth fishing before the tide is three and a half hours ebb. The boats set off at about 20 minute intervals, and if there are more than eight boats fishing, the last will not have time to drift the full length of the estuary before the flood begins to make. First boat to arrive at Baithaven Buoy is, by long-established custom, first away, and there is much competition to get down to the buoy before others. For this reason boats have been leaving their moorings earlier and earlier, with a long wait at the buoy in consequence before starting to fish. The race is always won by the men from Sunderland Point because they are nearest to the buoy and can see the others coming.

'But when you fish two tides in 24 hours, and each tide takes seven hours boating, the earlier you have to go to Baithaven Buoy the less rest you get. The first boat away from Baithaven probably stands the best chance of a good catch since there will be time for it to net the Bar pool at the end of the estuary twice. Watched by the others waiting at the buoy, the net is paid out and the man at the oars pulls over towards Cockersand Abbey Lighthouse. Down they go on a fast ebb, the corks disappearing out of sight in the waves, taking in a length of net past the Baulk, paying out again as they pass Smith's Skeer. The net has by now been swept into a long curve and is only fishing half its length. They row on to the far bank, then back again, paying out the net. This time they run the Bar, the water tearing down over the shallows taking net and boat in an uncontrollable flurry. Slowly they run up the back-water that takes them to the Bar again, while the boat that followed them down the estuary goes past in the full current. They fish out the tide, running the Bar pool once again, then wait for the flood to make.

'As the water deepens and the squalls drop into a steady breeze, they hoist sail and head the boat for home.'

A certain amount of drift netting for migratory trout is carried out off the Norfolk and Suffolk coast by no more than

11 teams of netsmen. In the area there are no longshore fishermen exclusively engaged in the netting of sea trout. There are, however, a small number of professional fishermen engaged in white fish, crab, lobster and shell fisheries operations, who seasonally augment their livelihood by the licensed netting of sea trout.[8] Most of the fishermen operate occasionally from such villages as Brancaster, Caister, Hunstanton, Wells, Runton and Winterton but no more than three boats fish from any one village.

In south-west England a certain amount of drift netting is carried out in the estuary of the River Camel and in the sea beyond. The quantities of salmon caught, however, is very low, the figures being 9 in 1966, 15 in 1967, 3 in 1968, 5 in 1969 and 3 in 1970. In 1966 27 migratory trout were caught, 17 in 1967, 7 in 1968, none in 1969 and 21 in 1970. Each drift net is 200 yards long, 10 yards deep, with a 1½ inch mesh.

A wide river with a strong tidal flow is essential for the efficient performance of a drift net, and so that as far as Wales is concerned it has a limited distribution. The principal drift netting areas of Wales are:

1. The Wye estuary where 2 netsmen,[9] employed by the Wye River Authority, operate gill nets, known locally as 'tuck nets'. Although the fishermen are based on the Authority's fish house at Stuart House in Chepstow, actual fishing is carried out in the Bristol Channel, beyond the estuary of the Wye.

2. The Usk estuary and Bristol Channel beyond where 8 boats, based on Newport docks, each manned by 2 fishermen, are engaged in 'tuck netting'. Most of the fishermen are members of an old-established Newport family.[10]

3. The Clwyd estuary in north Wales, where 8 boats each manned by 2 fishermen operate the so-called 'sling nets' from the coastal resort of Rhyl.[11]

4. The Dee 'trammel net' is somewhat different from the ordinary drift net, for although the method of using it is similar, the trammel itself is not a gill net, but is a complicated tangle net with two or three parallel walls of different sizes of mesh. 'It would appear that trammels are better fishing instruments than gill nets, because they are capable of taking a much larger

size range of fish, but to offset this they are much more difficult to use and where the currents are confused they may be impossible, for they twist and tangle into a hopeless mess.'[12] The Dee drifted trammel is a unique fishing instrument, and is today operated by only 4 2-man boats, all the fishermen being members of one Flint family.[13]

In addition, in Milford Haven, some of the compass netsmen of Llangwm and Hook are concerned with herring drifting, as well as drifting for other fish, such as bass, in the Haven, before the beginning of the salmon fishing season, that in practice commences in early June. They use the same 14 foot tarred boats as for compass netting, but use fine-meshed herring drift nets. This method of fishing, which strictly speaking falls in the category of in-shore fishing, will not be considered in this chapter.

THE WYE DRIFT NETS

The Wye estuary has long been regarded as one of the most productive stretches of river in Britain. Fishermen based on Chepstow, using a variety of instruments ranging from lave nets to basket traps and from stopping boats to drift nets, catch appreciable quantities of large-sized salmon. In 1968, for example, the single drift net, 4 stop nets and 700 putchers operated in that year by the employees of the Wye River Authority caught a total of 783 salmon, averaging 8.33lb in weight. Of the total number caught, 53.5% were caught with the drift net, 35.2% by stop nets and 11.3% by the putchers. The drift netsmen of the Wye, unlike those of other rivers, rarely fish for more than four hours per session. Nevertheless, it is unlikely that the use of drift nets in the Wye estuary is of any great antiquity, for although all mid-nineteenth-century reports mention stop nets, putts and putchers as being very common in the Wye estuary, nowhere are drift nets mentioned. It is probable that drift netting commenced during the first decade of the twentieth century, when the Conservators of the Wye Fisheries Association assumed responsibility for all netting on the Wye. In 1902, the Conservators passed a bye-law forbidding the use of drift nets in the non-tidal reaches of the river and secured a lease from the Crown of all the netting in tidal waters. This not only included the Wye itself, but that area in the Severn beyond the Wye estuary itself, that came under the jurisdiction of the Wye Conservators. Between 1902 and 1904 all netting in the Wye was suspended and was resumed in 1905 'on the very limited scale at which it now remains'.[14] It is probable that drift netting was introduced at that time, for although no more than half a dozen boats were engaged in drift netting during any one season, appreciable quantities of salmon were caught by the netsmen. Local tradition says that drift netting was introduced into the area by fishermen who operated on the Lune in Lancashire and who had settled in Monmouthshire during the first decade of the twentieth century. On the adjoining Usk fishing grounds, however, drift netting was not allowed until 1914 and was carried out largely by fishermen who migrated to Newport from the Bridgwater district of Somerset. By 1939 only three drift-netting boats were in use on the Wye and by 1965 only two remained. In 1971, one boat manned by two fishermen was engaged in full-time fishing during a season that extends from 2 February to 15 August. Most of the fish are caught during the months of May and June for, according to the fishermen, there is a marked decline in catches during July and August. Unlike the drift netsmen of any other river in Wales, the Wye netsmen only operate for four hours at a time, and fishing sessions of twelve or fourteen hours, so common on the Usk, Clwyd and Dee, are unknown on the Wye.

Although in the past boats equipped with a single lug-sail were widely used for drift netting, today a diesel-powered, 22 foot carvel-built vessel is used. The boat was built at Appledore, Devon, in 1965. The actual fishing grounds are in the wide River Severn, for the Wye itself is too narrow for the efficient performance of a net 400 yards or more in length. With the ebb tide, the netsmen drift from the mouth of the Wye as far west as Sudbrook, returning from Sudbrook to the Wye with the flood. There is no real limitation on the area of water fished, as long as it is the stretch of water designated as being under the Wye River Authority's jurisdiction; indeed fishing may be carried out as far south as the mouth of the Bristol Avon.

The nets used by the Wye drift netsmen are 400 yards long and 2 yards deep and today most are made of synthetic fibres. Until the late nineteen-sixties, however, cotton and hemp nets were usual and these had to be boiled at regular two-weekly intervals in a special boiler on the river bank at Chepstow. This boiler was to remove sludge, slime and salt from the nets. Cutch[15] was the most widely used substance for treating nets; the process of cutching a net 'consists of either dipping it two or three times in the hot solution of cutch water and drying it each time, or of soaking it for a considerable period, say two days. After the first cutching of a new net, it is also dipped periodically in the solution . . . cutch is the basis . . . for all drift net dressings as it clears out the vegetable oils etc from the twine'.[16]

The Wye River Authority specifies a mesh size of '2¼ inches from knot to knot or 9 inches round the four sides', but in practice nets with a mesh that varies from 3½ inches to 5 inches are used, at different periods of the fishing season. A small meshed net is used for catching grilse, early in the season, while a coarser meshed net is required for catching the large fish that are more common, late in the season. It is customary, therefore, for the River Authority to issue nets of a different mesh to their fishermen as the season progresses.

Although the Authority's bye-laws of 1953[17] state in the Schedule that nets may be armoured or unarmoured, the armoured drift net is not used on the Wye; indeed there is no evidence to suggest that armoured nets have ever been in use on the river, and fishermen have expressed the opinion that this type of net is illegal. The drift net is equipped with two 'rock staffs', each 4 feet long, weighted with lead at the bottom in order to keep the net upright in the water when fishing. In practice, the inner rock staff is discarded and the net is allowed to drift as the tide and wind will carry it, without being connected to the boat in any way. With an inboard engine it is relatively easy to keep up with the movement of the net. The fishermen regard a freely drifting net as being a far more efficient fishing instrument than a net connected to a rock-staff, which in turn is attached to a length of rope, up to 20 yards in length, attached to the boat. The net is hauled into the boat, usually at the end of the drift and the turn of the tide for the removal of fish, before it is re-set for the return drift to the Wye.

THE USK DRIFT NETS

Like the Wye, the Usk has long been regarded as a very important salmon river, and although putts, putchers and stop nets were widely used in the estuary and the adjacent parts of the Bristol Channel in the nineteenth century, there is no evidence to suggest that drift nets were used for salmon fishing until the early years of the present century. A member of the Sully family, who have been important in the fishing industry on the Usk, came to Newport from Somerset during the first decade of the present century and began drift netting in the Bristol Channel at that time.

Today, there are 8 licensees allowed to use 'tuck nets' in the Usk River Authority's waters,[18] although they are barred from fishing in the estuary of the Usk itself by an Act of Parliament in 1911. They are limited in their fishing to an area south 'of an imaginary straight line commencing at Goldcliff lighthouse and passing through No 1 buoy . . . and continuing until it meets the coastline at the point indicated by a marker 15 feet high . . . on the side facing Goldcliff'.[19] The in-shore limit of drift netting is therefore along an approximate line drawn from St Brides Wentloog to Goldcliff, while the western limit of fishing is in 'those tidal waters which lie to the east of a line drawn in a 43½ east of south true direction, from a point at high water mark approximately one furlong west of Collister Pill'. Until 1939 two boats based on Newport were licensed by the Usk Board of Conservators to fish in the Rhymney estuary near Cardiff, and there was considerable controversy as those fishermen often set their nets on leaving the Usk and drifted to their allotted fishing grounds, catching salmon in the Usk waters as they drifted. Until 1936, when the Usk and Rhymney fisheries districts were amalgamated, there were four teams of drift netsmen licensed to fish in the Rhymney area. Fishing usually finishes off the coast at Peterstone Wentloog, though drifting is allowed as far as the mouth of the Rhymney. Today the fishermen may fish eastwards as far as the Severn tunnel, the waters beyond Sudbrook being under the jurisdiction of the Wye River Authority. No nets or fixed engines of any kind are used on the Usk itself, although both the Goldcliff and Porton putcher ranks come under the jurisdiction of the Usk River Authority. Until the early nineteen-thirties stop nets were allowed in the Usk.

Until the nineteen-thirties, too, armoured nets with two walls with 4 inch and 18 inch meshes were used for salmon drift netting from Newport, and although trammels were used in recent years for flat-fish catching, trammel nets are illegal for salmon fishing. The only nets allowed by law are 'Drift nets . . . without bags or pockets and without armour. Such nets shall consist of a single sheet or wall of netting, measuring not more than 9 feet in depth and having a mesh of not less than 2 inches from knot to knot or 8 inches round the four sides. The total length of drift net used by any one boat shall not exceed 300 yards in length.'[20]

According to the annual reports of the Usk Board of Conservators, it was only in 1937 that the term drift nets replaced that of trammel nets and until the amalgamation of the Usk and Rhymney fisheries districts in 1936, both armoured and unarmoured nets were used in the areas. The catches and number of licensees allowed to operate in the river, from 1910 until the old Board of Conservators was ended in 1952, are shown in Table 10 *(below and page 211).*

TABLE 10
LICENSEES AND SALMON CATCHES BY TRAMMEL OR DRIFT NETS, USK, 1910-52

1910-13 No trammel or drift net licences issued for the river. All salmon caught by stop nets and putchers and putts

	LICENSEES	SALMON CATCH (No)
1914*	12	351
1915-16	No returns	No returns
1917	14	453

TABLE 10 *(continued)*

	LICENSEES		SALMON CATCH (No)	
1918	15		164	
1919	12		427	
1920	12		350	
1921	17		375	
1922	13		396	
1923	16		482	
1924	18		734	
1925*	30		1512	
1926	18		1124	
1927	No figures available			
1928	13 (Usk)	4 (Rhymney)	2076 (Usk)	319 (Rhymney)
1929	18 (Usk)	4 (Rhymney)	921 (Usk)	133 (Rhymney)
1930	18 (Usk)	4 (Rhymney)	745 (Usk)	111 (Rhymney)
1931	18 (Usk)	4 (Rhymney)	1875 (Usk)	280 (Rhymney)
1932	18 (Usk)	3 (Rhymney)	2398 (Usk)	276 (Rhymney)
1933	18 (Usk)	3 (Rhymney)	2569 (Usk)	232 (Rhymney)
1934	18 (Usk)	3 (Rhymney)	2515 (Usk)	172 (Rhymney)
1935	18 (Usk)	2 (Rhymney)	2080 (Usk)	156 (Rhymney)
1936	18 (Usk)	2 (Rhymney)	1631 (Usk)	139 (Rhymney)
1937	20 (Usk and Rhymney)		1349	
1938	18		602	
1939	19		1249	
1940	16		977	
1941	15		999	
1942	16		821	
1943	16		620	
1944	16		377	
1945	16		803	
1946	20		759	
1947	20		927	
1948	20		1309	
1949	20		1053	
1950	20		697	
1951	20		790	
1952	16		1067	

*Trammel nets were approved for the first time by the Board of Agriculture and Fisheries in 1914.

On 22 January 1925 a resolution to limit the number of trammel netsmen to 18 was made by the Usk Board of Conservators.

Each boat employed in drift netting is manned by two men and each licensee, paying an annual fee of £35 in 1971, may endorse two other names on his licence. There has been a decline in the number of netsmen operating from the Usk, for in 1930, when there was no limitation on the total number fishing, 30 boats were engaged in drift netting. By 1939 a limit of 18 boats was instituted by the authorities, by 1946 this had been limited to 12 and in 1948 the number was fixed at the present limit of 8. A licence, if given up for some reason by a fisherman, can be taken up by another, usually by someone whose name is endorsed on the licence. There must, however, be no more than a total of eight boats in operation.

In using drift nets, 17 foot, grey-coloured, clinker-built boats, with 6 foot beams, are used. To reach the fishing grounds a 4hp inboard engine is used, but during actual fishing the engine is not used and the boats drift freely with the tide. The fishing season extends from 2 March to 31 August, and each session in the Bristol Channel extends for twelve hours or more. In addition it usually takes an hour and a half to reach the fishing grounds from the Pier Head at Newport and another hour and a half to return from the fishing grounds. Until 1930 each fishing boat was equipped with a single lugsail, and it was only in 1930 that petrol engines made an appearance in the Usk fishing fleet. Boats have always been stable and sturdy, most of those in use today being built by a Caerleon boatyard to the specification of the fishermen themselves.

Each net, which is 300 yards long and 3 yards deep, is equipped with a pair of leaded staffs, each 4 feet long. The net is corked along the top and leaded along the bottom, so that it floats vertically in the water. It is connected to the boat by a rope approximately 10 yards in length and each net has to carry a licence number attached to the corked head-rope. 'The same number preceded by the letter U shall be conspicuously painted and maintained on the outside of any boat or vessel, from, or in connection with which the licence is granted.'[21] In addition a pole, 3 feet long and carrying a fluorescent orange flag, has to be attached to the outer staff of the net.

To use a drift net, 'One end of a drift net shall be attached to a leaded end-staff or pole and the other end of the drift net shall be attached to a boat, which shall be manned by not more than two persons. The net shall be shot or paid out from the boat, which shall be fastened to one end only of the net by a rope and the net shall be used by floating or drifting with the tide. No boat shall carry at any time a total length of net or nets exceeding 300 yards.' In the 1930s drift nets of the trammel variety were limited to a length of 100 yards, but the 1934 Report of the Usk Board of Conservators says that most boats 'take additional nets while they join up so as to considerably increase the 100 yard limit'. In drifting it is important that the net is kept out straight against the tide, and this involves the constant use of a pair of oars. Care must also be taken to ensure that the net is not too taut, or this will cause rotting and tangling.

Appreciable quantities of salmon are caught by drift nets. Between 1960 and 1970 the following quantities were caught:

1960 – 562
1961 – 466
1962 – 737
1963 – 647
1964 – 956
1965 – 876
1966 – 775
1967 – 631
1968 – 621
1969 – 1063
1970 – 1004

In 1971 the catch was as follows:

	Salmon	Sea Trout
March	5	1
April	15	4
May	139	7
June	265	-
July	239	-
August	103	-
Total	766	12

Again, varying meshes are used for, although the minimum specified by law is 2 inches, the deciding factor is the size of fish the fishermen believe to be in the fishing grounds at any particular time. If they think there is a run of large summer fish they use a 3¾ inch mesh net. The net is piled on to the bows of the boat and paid out as the boat moves. The aim is to keep the net as straight as possible, so that it drifts with the tide at its maximum length, gill-meshing the salmon. Changes in the speed of the currents, cross eddies and varying depths soon put kinks in the net, so that the drift net is not always used at its full length of 300 yards.

CLWYD SLING NETS

In May 1866, a Board of Conservators for the Clwyd and Elwy was formed and their first task was to make illegal the double trammel nets that had been widely used in the tidal reaches of the Clwyd below Rhuddlan Bridge. Undoubtedly the trammel, which was in all probability similar to that still used on the Dee, was by far the most common fishing instrument on the Clwyd, for 'the three lower miles of the joint rivers were open to public netters, who fished fourteen or fifteen double trammel nets, each worked by four men; these nets were formed by a centre one of a small mesh on either side of which hung a "wall" of net with a large mesh; when a fish struck the centre net, its weight carried a portion of it through the large mesh net on one side or the other and thus the fish bagged itself. By the Act of 1861 these nets, fortunately for the fish, became illegal, though it was not till a few years later that the fishers brought themselves to recognise the law'.[22] The *Second Report of the Inspectors of Salmon Fisheries*(1862)[23] advocated this move and suggested that seine nets should replace the traditional double trammel. Earlier in the nineteenth century, it is interesting to note that a number of salmon fishermen from the Solway set up a series of stake nets in the Clwyd estuary in 1840,[24] but 'the native fishermen took the law into their own hands, smashed up the stake nets and kicked the intruders out of the county'.[25]

By 1861, there is evidence to suggest that drift netting was already being practised on a limited scale, especially along the sea shore beyond the estuary of the Clwyd.[26] Hoylake boats paid periodic visits to Rhyl and fished for salmon and cod in the sea, while a fisherman from the town fished the estuary for 'salmon, buntling shrimps and cod'. The area that he fished extended as far as Rhuddlan Bridge, that is the tidal reaches of the Clwyd. With the banning of trammel nets in 1864, the number of both drift and seine netsmen increased considerably and appreciable quantities of salmon were caught throughout the nineteenth century. For example, in 1872, netsmen took 365 salmon during a season that extended from 1 March to 31 August. In 1874 they took 1,046 fish, 509 in 1875, 1,834 fish in 1876, 620 in 1877 and 1,070 in 1878. In 1879 the fishing season extended from 15 May to 15 September only and a total of 625 salmon was caught in that year. Between 1880 and 1902 the following were the quantities of salmon caught, together with sea trout, on the Clwyd: 1880 – 570, 1881 – not known, 1882 – 1,400, 1883 – 2,800, 1884 – 2,300, 1885 – 2,300, 1886 – 2,660, 1887 – 2,700, 1888 – 3,000, 1889 – not known, 1890 – 4,300, 1891 – 4,222, 1892 – 2,527, 1893 – 2,219, 1894 – 2,106, 1895 – 2,200, 1896 – 2,000, 1897 – 1,600, 1898 – 1,000, 1899 – 1,000, 1900 – 700, 1901 – 1,000, 1902 – 1,300.

Undoubtedly the heyday of salmon fishing on the Clwyd was between 1883 and 1896 when appreciable quantities of both salmon and sea trout, averaging 6½lb each, were taken by approximately 80 netsmen.

Since the eighteen-nineties there has been a decline both in the number of netsmen and in the quantities of fish caught in the Clwyd. In the years 1903 to 1909, for example, the total number netted was 6,000 fish, averaging 857 a year – a considerable decline from the 4,300 taken in the 1890 season.

Within recent years there has been a rapid decline in the number of drift netsmen operating from Rhyl. In 1950, 20 boats were engaged in drifting, but in 1970 only 8 licences were issued.[27] In addition, drift nets operated without a boat were commonly used by Clwyd fishermen. This method of fishing, known locally as *brewas*, involved the use of a drift net, attached to a long rope held by a fisherman on the shore. As the vertical net drifted with the tide, the netsmen kept pace with it by walking along the shore, so that the net was at right angles to the flow of the tide. This method of drifting has not been practised on the Clwyd for at least twenty years. Each of the eight netsmen operating from Rhyl today may endorse up to four people on a licence, although in practice only two fishermen are employed in a boat at any one time.

The fishing area extends from below the Foryd Railway Bridge across the River Clwyd near Rhyl and fishing is usually carried out both in the estuary and in the sea between the Point of Ayr and Llanddulas. The fishing season extends from 15 March to 31 August, although in practice few of the fishermen venture on the river before mid-June, the best catches being taken in July and August. The sling net consists of an unarmoured net 200 yards long, 5 yards deep and with meshes measuring no less than 2 inches from knot to knot. Clwyd sling nets, unlike those of the Usk and Wye, are not equipped with end-staffs, but in order to keep them upright in the water while fishing, they 'may be weighted at either end by a weight not exceeding 9lb'.[28] Half a house brick is the usual weight tied on to the lead line of a Clwyd sling net, which is usually 22 meshes deep, although nets up to 25 meshes deep may be used occasionally. In the nineteen-twenties, the hemp salmon drift nets used on the Clwyd were limited to a length of 120 yards with a depth of 30 meshes, the mesh being 3¼ inches.[29]

Fishing begins with the ebb tide and the fishermen in their 16 foot boat either drift downstream from the estuary, a process known as 'aiding down', to the open sea, or they use an outboard engine to take the boat to the Point of Ayr to begin fishing. No engine is needed during the actual fishing, for oars are usually sufficient to keep the net at right angles to the drifting boat. Each fishing session extends from ten to thirteen hours and both day and night tides are fished.

In 1970 the catch of fish was as follows:

	Number of salmon & sea trout	Weight (lb)
March	0	-
April	0	-
May	26	155¼
June	53	297¼
July	820	4071¾
August	363	2004½
Total:	1262	6528¾

THE DEE TRAMMEL NET

The stronghold of trammel netting has always been north-east Wales, for although trammelling was practised in the Bristol Channel and in the Menai Straits in the past, it was on the Dee and Clwyd that trammel drift nets reigned supreme. Trammel netting was banned by an Act of 1861, and by 1864 trammels had disappeared from the Clwyd, but 'on the Dee they were worked as if the Act of 1861 had not been passed'.[30] It was estimated that in 1862 a total of 160 men were regularly employed on trammel and draft netting on the Dee, while in addition there were '30 or 40 scowbankers, who, otherwise employed during the week, pay the estuary a visit on Saturday and Sunday, hire or borrow boats, and fish away to the great annoyance of the regular fishermen'.[31] It is impossible to say how many fishermen were employed in trammelling on the Dee before the eighteen-sixties, but one witness to the Commissioners on Salmon Fisheries in 1861[32] noted that, in the eighteen-twenties and thirties, there were at least ten boats trammelling on a full-time basis in the Dee estuary. By 1862 it was estimated that below Queensferry there were 40 boats employed in seine and trammel netting during a season that extended from 1 March to 31 August.[33] Nevertheless, little attention was paid to the limitation of season set down by the Dee Association. 'By Orders of the Association the bellman used to be sent to the fishing villages in the neighbourhood, and the close season was cried. The fishermen of these villages dropped down with the tide, and set to work again, beyond the clang of the bell and the ken of the Association.' Undoubtedly the Dee in the eighteen-sixties was being grossly over-fished throughout the year by netsmen, and as a result fixed nets and trammel nets were declared illegal and only 47 drift nets and 18 coracle nets were allowed on the river.[34] Nevertheless, despite legislation, trammel netting continued unabated throughout the sixties and seventies, and a large number of fishermen were prosecuted every year for using illegal trammel nets. In 1868, for example, 'twenty convictions for poaching were obtained, chiefly for spearing, for using the trammel net and for breach of the weekly close time or resisting the bailiffs'.[35] In 1878, however, trammel nets were again legalised, and in that year 7 licences at £10 each were issued. By 1888 the number of trammel net licences had been increased to 22, and, said one observer,[36] since trammel nets 'are a specially deadly net and illegal in all the rivers, the Conservators informed the Fishery Board that they proposed to

Trammel netsmen on the River Dee

abolish them; strange to relate, their proposal was not sanctioned, and so in order to keep down their numbers the licence for a trammel net was raised from £10 to £15'. In that year too, the fishing season was extended to cover the period 1 February to 31 August. By 1900, the number of trammel net licences had declined to 12, but there was no real limitation on number as long as licence fees were paid. In 1920, however, as a result of a disastrous season, a limit of 4 trammel licences was instituted. The limitation to four teams is still in force at present.

Although fixed trammels have been known in various parts of Britain such as the Somerset coast within recent years,[37] drifted trammels were and are known only on the Dee, where 'local regulations forbid the fixing of trammels, which must therefore be drifted, the set of the net being across the tide'.[38] The current Fishery Bye Laws[39] note 'a trammel net shall be shot or paid out from a boat which shall be manned by not more than two persons in all. One end of the net shall be fastened by a rope to the boat and the other end to a floating buoy or float, and the boat, float and net shall not be made stationary in any way, but shall be allowed to drift with the tide.'

The Dee trammel used by four teams of fishermen based on the town of Flint is a 100 yards long and consists of a central wall of netting known as the lint, with a mesh of 2¼ inches. The lint, which is loosely hung to the head- and foot-ropes, has a mesh too small to gill the fish, but attached to the ropes are two outer walls of much larger, square-shaped meshes, measuring 11 inches from knot to knot. These are called 'the armouring' and are not as slackly hung as the lint. All three walls are set on to a common rope at the top and bottom, the lint naturally hanging slackly between the armourings. The maximum height of the armourings is 51 inches each from corked head-rope to lead-lined foot-rope, but the lint is 30 inches more, being 81 inches from head-rope to foot-rope. Although the armouring on present day trammel nets is square, diamond mesh was often used in the past, but a number of Flint fishermen expressed the opinion that square armouring was more efficient. The lint of the net nowadays is bought from a large-scale Bridport

Pulling in a trammel net

net-maker and is usually of fine cotton or hemp. The armouring is braided by the fishermen themselves, and although hemp or cotton twine is usually used for this now, in the nineteen-thirties, carpet thread was usually used for the armouring.

The principle of the trammel is that when the net is set across the flow of a stream and the probable direction of the movement of fish, any salmon that swims against the net passes through the armourings nearest to it. It 'strikes the slackly hanging lint, and by its impetus carries a bag of it through a mesh of the further armouring. Thus the fish is caught in a bag, the mouth of which is constricted by the mesh of the armouring through which it has just been pushed, and the net is equally effective whichever side the fish strikes'.[40]

The boats used by the Dee trammel netsmen are clinker-built 18 foot vessels, equipped with an inboard engine and a transom half-deck for carrying the net at the stern. The boats are built to a traditional pattern by a Chester boat-builder[41] and the moulds for the Dee trammel boats have been kept by the boat-builder for at least fifty years, so that any new boat required is built to the old pattern. The design of the boat is especially adapted to the character of the river, for each one is 6 feet 1 inch – no more and no less – in beam, and is sturdy and well-balanced enough to withstand the considerable buffeting of a rapidly moving tide.

The four licensees on the Dee can each endorse up to three other names on their licences.[42] The most important family in the trammel netting business is the Bithell family, who figure in Flint Church Records for nearly seven centuries, and who according to family tradition have been concerned with trammelling from at least 1600. Most members of the family live close to the banks of the Dee at Flint and fishing is carried out from a buoy between Flint and Queensferry and another buoy near Mostyn, nine miles down river. The whole course of the river and its rapidly flowing channels, known locally as 'swims', is marked by a series of ten buoys, each one a mile from the next. Usually fishing commences at Buoy Number 1, and an inboard engine, with which each boat is equipped, is used to propel the boat to the beginning of its fishing. With a rapidly flowing ebb tide, it may take a little time to reach the first buoy. The engine is then switched off and the fisherman depends on a pair of oars to steer the boat to the correct position for casting the net. This is done by the fisherman's assistant who stands on the half deck at the stern of the boat, casting the net in such a way that it is at right angles to the rapidly drifting boat and the flow of the tide. A buoy, usually in the form of a plastic float or plastic bottle, is attached to the outer end of the net and this gives an indication of the position of the net in relation to the boat at any one time. Should the net not be at right angles to the boat, as often happens when there is a variation in tidal flow in the swims, the net has to be pulled in and cast again. For night time fishing, the plastic float is replaced by a light, which acts as a guide. Drifting with the tide demands the considerable use of oars to keep the boat on the correct line. Usually drifting is continued until Buoy Number 10 has been reached, and with the flood tide it continues again until the boat is back at Buoy Number 1. The stretch of river above that buoy is fished by the Queensferry seine netsmen.

The salmon fishing season on the Dee extends from 1 March to 31 August, and the Bithell family being full-time fishermen spend the rest of the year trawling for flat fish and shrimping and indeed, should the salmon season be a particularly poor one as the 1970 and 1971 seasons were, the fishermen may return to shrimping even in summer.

In 1970 the catch of salmon was as follows:

	No	Weight
March	0	0
April	2	23
May	3	36
June	7	53
July	228	1329½
August	118	777½
Total:	358	2219lbs

For notes to Chapter 8, see pages 319-20

9 Seine Netting

The shore seine[1] is widely used throughout Britain, more especially in river estuaries, and it is by far the most common instrument for the capture of migratory fish. The net is a simply constructed, plain wall of netting, 200 yards or more in length and of a depth suitable to the water in which it is used. It is important that the net should extend as far as possible from the surface of the water to the bottom, and should stand as vertically as possible in the water. The head-rope is fitted with cork or plastic floats and the foot-rope weighted with lead or stones. The net is carefully stowed on the flat transom of a small boat. One of the crew stands ashore, holding a rope attached to the end of the net; the boat is then rowed out from the shore on a semicircular course, with the net being paid out over the stern. When the whole net has been shot, the boat returns to the shore whence it set out. The crew then lands, the boat is made fast and the net is hauled in. The landing place is usually downstream of the shoreman, but the net is occasionally shot upstream, particularly if the boat, as on the Teifi and Dee for example, is equipped with an outboard motor. The hauling of the net is rapid and smooth, and the two ends of the net are brought close together, 'thus making a narrow bag of the middle of the enclosed space, where the fish are concentrated and can be hauled ashore. The foot rope is hauled faster than the head-line, thus making a more pronounced bag of the centre of the net'.[2] The whole net is then drawn in; the salmon or sea trout caught in its mesh are killed with a wooden knocker, and the net has to be rearranged on the boat transom, ready for the next cast.

In 1970, the following seine nets were licensed by the various river authorities in England and Wales:[3]

RIVER AUTHORITY	NUMBER OF LICENSED NETS
Cumberland	Esk – 1 Eden – 3
Lancashire	Lune – 1 Duddon – 2
Dee and Clwyd	Dee – 31
Gwynedd	Conway – 6 Dyfi and Dysynni – 8 Mawddach – 3 Glaslyn – 3 Dwyfor – 2 Daron – 1 Seiont – 4 Anglesey – 2 North Caernarvonshire – 1 Ogwen and Aber – 2
South-West Wales	Tywi – 9 Taf – 1 (Wade Net) Teifi – 6 Cleddau – 4 Coastal nets – 3
Severn	Severn – 4
Devon	Axe – 1 Avon – 1 Teign – 10 Dart – 18 Taw and Torridge – 36
Cornwall	Tavy – 3 Tavy-Lynher – 2 Tamar – 14 Tarmar-Tavy – 5 Lynher – 3 Fowey – 4 Looe – 1
Avon & Dorset	Avon, Stour, Frome, Piddle – 30

SEINE NETTING IN WELSH RIVERS

In some estuaries, such as those of the Wye and Usk, the muddy and steep character of the river banks is such that seine netting is impossible, but in most other estuaries, from the Dee in the north to the Tywi in the south, the seine net reigns supreme. In 1971, there were 86 licensed seine nets in Wales. On the Dee, for example, 30 boats employing 72 fishermen are licensed to net salmon on the river below Old Dee Bridge at Chester, most of the fishermen being based on Chester and on Connah's Quay in Flintshire. On the Teifi, the Seine Net Fishermen's Association at St Dogmaels, Cardiganshire, has 24 members who fish from half a dozen boats in the estuary while on the Tywi there are 9 seine net boats based on the village of Ferryside, Carmarthenshire.

Despite its occurrence in estuarine waters throughout Wales, there are local variations in the dimensions and techniques of using the seine net. On the Teifi, for example, nets are usually

180 to 200 yards long and 3 yards deep with a 2 inch mesh.[4] On the Tywi a 1½ inch mesh is allowed. The Gwynedd River Authority, responsible for such rivers as the Conway, Mawddach and Dyfi, specifies seine nets not more than 150 yards long and 6 yards deep, but the mesh of net is not specified. On the Dee, on the other hand, nets may be up to 600 yards long, 5 yards deep with a mesh of not less than 2 inches.[5] In south-east Wales, seine nets were used on the Wye, above Tintern, in the nineteenth century, hand-operated winches being fixed to the river banks for hauling in the nets, but nowadays no seine nets are used on the Wye or Usk or any Welsh river further east than the Tywi. In Gloucestershire, seine nets known locally as 'long nets' are used for salmon catching on the deep water reaches of the Severn, below Tewkesbury. In 1970 there were four licensees. 'The long net', says Taylor,[6] '. . . is shot from a flat-bottomed Severn punt. The lines of one end of the net are left with a "debot-man" on the bank while the net is paid out or "shot" from the punt as it crosses to the opposite bank. The punt then turns downstream until the whole net has been shot. The lines called "the muntle" at this end are then thrown back across the river to the muntle-man waiting on the bank and the haul begins. By ancient custom, the debot-man may claim any fish other than salmon caught in the net . . . '

TEIFI

Draft nets are used in the Teifi estuary below the village of St Dogmaels for catching salmon and sea trout, during a season that extends from 1 March to 31 August. According to the *Report of the Commissioners* in 1861, 42 seine nets were used by 16 teams, consisting of 'worn out sailors and boys, who have not gone to sea'. Upstream in the non-tidal reaches of the Teifi, two seine nets were used at Llechryd and another at Cilgerran. The *rhwydau sân*, as they are known locally, measure up to 200 yards in length, have a mesh of 2 inches from knot to knot and are from 10 feet to 12 feet in depth. Rowing boats equipped with outboard motors are used by the Teifi seine netters, boats that in the past were always referred to as *llestri sân* (seine vessels). Until 1935, when the first outboard engine was introduced into the seine boats, black-tarred, 20 foot rowing boats were used exclusively for seine netting. Before 1920 the boats were usually 25 foot rowing boats equipped with curved oars and without rowlocks. The oars were designed to go through holes in the sides of the boats and the boats ended their lives by being concerned with carrying cargoes of gravel on the river. In those days, it was customary to venture beyond the 'bar' at the mouth of the Teifi into the open sea beyond and nets of the full length of 200 yards were required. Today however, since most of the fishing takes place in the river itself, a net of 150 yards or so is usually long enough for salmon fishing.

Undoubtedly seine netting has been well known on the Teifi for many centuries. George Owen, writing in 1603 for example,[7] describes the 'great store of salmon as allso of sueings, mullettes and botchers' taken in the Teifi 'neere St dogmells in a sayne net after every tyde'. According to the Inspectors of Salmon Fisheries in 1862[8] there were 16 draft nets operating in the river, while in addition there were a number of fixed nets, known as 'jackass nets', in use near Cardigan Bridge. The most interesting of the fixed nets, however, was the unique net known as '*y shot fawr*' fixed across the estuary of the Teifi, near the bar by the seine netsmen. In 1895, the 'great net' *(shot fawr)* used at the mouth of the Teifi was declared illegal but, for a few years after, it operated in Milford Haven. The '*shot fawr*' consisted of two draft nets tied together and placed across the flow of the river. 'It had long been the custom of draught-net fishermen in the public waters below Cardigan', says one observer,[9] 'to take it in turn to join two of their nets together and to use the long net thus formed in a different way to the ordinary manner in which the other draught nets were used. In this mode of fishing, each of the twenty draught nets in the estuary would get its chance of becoming the "*shot fawr*" once in every five days. The two nets when joined made a length of nearly four hundred yards, which was stretched and fixed across the top of the estuary, and so every salmon that fell back with the tide was caught by it, and as many as 170 had been taken at one shot. In times of continued drought, this net was therefore certain to capture the whole of the fish that had gathered together when waiting for a flood.' It was maintained by anglers and coracle fishermen that the presence of the great net in the estuary of the river was damaging the supply of sea trout and salmon and after much discussion the 'Shot Fawr' was eventually doomed and ceased to exist in 1895. An earlier description of *y shot fawr* in the *Second Annual Report of the Inspectors of Salmon Fisheries*[10] in 1862 describes the net as being '440 yards long . . . set in the estuary during the last two or three hours of ebb tide and takes every fish returning with the ebb to the sea'. 'The portion of the river where it is used is public', says the Report, 'but by a custom among the fishermen those men only have a share of the "shotfawr" . . . who attend, on a certain appointed day at the commencement of each season, a meeting called "Cwmgwyr" held for this purpose, when partners are chosen and each fisherman contributes his share of netting, to be formed into one common net. There are 14 men in all and each man, or each man with his partner, has in turn the proceeds of a tide; such assistance as is required for working the net being rendered by the others to the boat whose turn it is to take the fish.' In 1861 the Teifi seine netsmen were summoned for using a fixed engine, but the plea of 'immemorial usage' was set up and the case was dismissed.

During the course of the present century, there has been a decline in the number of seine net fishermen operating in the Teifi estuary. In 1939, for example, there were 13 boats, each one manned by a team of 5 fishermen operating in the Teifi; today the number has decreased to 6 boats, each boat manned by 4 men only. The owner of the boat and nets is always described as *Y Capten* (the Captain) and his crew as *gweision* (servants, singular: *gwas*). In the nineteen-twenties, each team consisted of 7 fishermen, and it has been estimated that 20 boats were engaged in seine net fishing at the time, compared with 23 boats in 1900, operating throughout a season that extended from 16 February to 31 August. Today the fishing season commences on 1 March and the river is closed between

Seine netting in the Teifi estuary

6.00am on a Saturday to midday on the following Monday. The 24 full-time fishermen of St Dogmaels are all members of the St Dogmaels Seine Net Fishermen's Association, whose headquarters is at the Teifi Inn at the Netpool in the village.[11]

There are a number of reasons for the decline of seine net fishing in the estuary of the Teifi. The fishermen themselves tend to blame the River Authority for severely curtailing their activities, limiting the season to six months of the year and limiting the size and mesh of the nets. High licence fees of £28 per net per annum are also blamed for the decline in the number of fishermen, the licence fee at the beginning of the present century being no more than 32s 6d per boat. Undoubtedly one of the main reasons for the decline has been the gradual silting up of the estuary and a change in the course of the river. Silting has followed the decline of Cardigan as a seaport, for when the wharves of Cardigan flourished, a navigable channel was kept open to the sea. Sand from the estuary was also used as ballast for the ships. As a result of the decline of sea trade, the river has silted up, and there would literally be no room for the 23 seine nets of less than a century ago. Pools that yielded appreciable quantities of salmon have silted up and the seine netting today is concentrated in four pools only, *Pwll Nawpis, Pwll-y-Perch*, and *Y Gwddwg* beyond *Pwll-y-Castell* and *Pwll Sama.* Until 1939, other pools, such as *Pwll Nhwicyn, Pwll Nant-y-Ferwig, Pwll-cam, Pwll Pric, Pwll Parchus, Pwll-y-Brig, Pwll Rhipin Coch* and *Pwll Wil-y-Gof,* were all well-known salmon pools, but have by now virtually disappeared.

To decide which team of seine netsmen goes to which stretch of river on a particular day or for two tides, lots are drawn daily at the Netpool, by picking numbered stones (*cymryd shot* – to take a shot) from a bag or hat.[12] Every boat must, however, take its turn at Pwll-y-Castell, the nearest fishing pool to St Dogmaels village. To decide the order of fishing at Pwll-y-Castell, lots are drawn on the first day of the fishing season. The team drawing disc Number One must go to Pwll-y-Castell on the first day of the fishing season; that drawing Number Two goes there on the second day; Number Three on the third day and so on in rotation throughout the season. To make this draw for the right to fish in Pwll-y-Castell, it is customary for the leader of the team that took out a licence last to hold the bag or hat containing the numbered discs for this draw. If for some reason only five boats have licensed their nets for fishing on the first day of the season, the sixth boat has to go to Pwll-y-Castell on the first day it is out fishing. It then takes its place in the rotation.

With the exception of the boat and its team that is destined for Pwll-y-Castell for two tides on any one day, lots have to be drawn daily to decide where the other five boats have to fish on that day. The last team leader to reach the Netpool on the previous night has the right to hold the bag or cap containing the numbered discs, while the first boat to land at the Netpool the previous night has the right to have the first pick of the stones. The team drawing Number Five has to go to Pwll Sama, located in a narrow stretch of the estuary between sand banks and where the depth of water at low tide is usually more than 14 feet. The team drawing disc Number Four has a choice of fishing at either Pwll Nawpis below Pwll-y-Castell or it can go downstream for another mile or so to Y Rhyd, a stretch of river marked by a tree trunk in the middle of the river between Pwll Sama and Pwll y Perch. Boats drawing Numbers One, Two and Three have also to go to Y Rhyd and Pwll y Perch and a stretch of river known as Y Groyn below. Numbers One and Two, however, have the choice of either fishing in Y Rhyd or they can go even further downstream to the mouth of the river behind the bar to a fishing station known as 'Y Gwddwg' (the Neck). If those two boats wish, they can go beyond the estuary northwards to net in the open sea from the beach. This area is known as 'Amgnoi' while the area to the south of the estuary, hardly ever used for fishing today, is known as 'Amgreinio'. If the two boats are fishing at any time in the sea below the cliffs at Gwbert, the boats that draw Numbers Three and Four may fish in the Groyn and Y Gwddwg behind the sand bar at the mouth of the river.

Before 1939, when there were 13 teams of seine netsmen in the Teifi estuary, the method of drawing lots for Pwll-y-Castell was similar to what it is today. Number Three boat would go to Pwll Nawpis, Number Four to Nant-y-ferwig (now silted), Number Five to Pwll Sama and beyond, whilst Numbers One and Two could go over the bar to the sea. The other teams could also fish in pools not taken up by boats allotted to Pwll-y-Castell, Nawpis, Sama and Nant-y-ferwig and could go over the bar. Numbers Three and Four were allowed in the open sea, especially if the tide in their allotted pools was below 12 feet in height. Once a boat passed its allotted pool downstream, it could not return to that pool on that tide and it was considered that it had lost its rights to that pool.

The Teifi seine net is unusual, for although it consists of a single wall of netting 150 to 200 yards in length, it is made up of seven nets attached to one another, each net having a specific name. In the centre of the trawl of nets *(y trân)* is the strongest net with a 2 inch mesh *(llygad)* made from twine that is twice as thick as the remainder of the *trân*. This is known as *Y Rhwyd Gôt* and is approximately 20 yards long; the usual method of describing its size being 10 to 12 *gwrhyd*.[13] On either side of *y rhwyd gôt* are three pairs of nets each 14 *gwrhyds* in length.

They are *rhwyd nesa'r gôt, rhwyd ganol* and *rhwyd fastwn;* the mesh on each one of them being slightly larger at 2½ inches from knot to knot. The cotton twine or nylon in the three pairs of outer nets are less thick, so that the whole net is slightly lighter to haul than if the whole net were made of the thicker twine as in the *rhwyd gôt.*

When cotton and hemp ropes were used on the river, it was customary to buy two new Bridport-produced nets every year, in addition to a new central net. The two new nets were inserted on either side of the central net in the *rhwyd nesa'r gôt* position, the outer net or *rhwyd fastwn* being discarded at the end of each fishing season. In this way a length of net would be used in one of the positions on *y trân* for three fishing seasons; that is, for the first season it would act as *rhwyd nesa'r gôt;* for the second as a *rhwyd ganol* and for a final season as a *rhwyd fastwn.* Nets made from synthetic materials last longer than those of cotton. In using a seine net, the shoreman *(bastwynwr)* holds on to a short line *(rhaff lan* or *y fals)* approximately 12 feet long which is attached to the head-rope of the net. Although an anchor may help the shoreman to hold on to the end of the rope, especially in fishing positions near the bar of the river where the tide is especially strong, in the past a pole *(y bastwn)* similar to those used in north Wales rivers such as the Glaslyn and Seiont was regarded as an essential piece of equipment for the shoreman. This was of oak, 6 feet long and 2 inches in diameter, and its tip could be driven into the sand if the tide was especially strong. Today poles are not used on Teifi seine nets although the terms *bastwynwr* and *rhwyd fastwn* have persisted in the terminology as a reminder of the days when a pole *(bastwn)* was used on all seine nets.

A seine net is equipped with a cork-lined head-rope *(y tanne)* and a lead-lined bottom rope *(windraff* or *godre).* The meshes of a seine net hang diamondwise, and for a net to hang thus, three meshes should occupy the lateral space of two fully stretched meshes. This method is known as 'setting in by the third' and it is known on the Teifi as *traeanu*. The method of setting was described as follows by one informant:[14] 'There are twelve eyes (ie meshes) to each traeanad. Measure 6-7 eyes along the length of rope. Mark the rope with chalk on either side of the 7 eyes. Take 12 eyes and tie each one to the headrope between the two chalk marks. After grasping the first 7 eyes, prepare a stick equal to the length of the 7 eyes together. This is then used to measure the length of each *traean*.'

The various sections of head-rope along the top of the net are tied together with a 'holdfast knot', but the sections of lead-line are not knotted as this would cause the net to roll when in use. The sections of lead-line, one for each of the seven nets, have to be bound together with cotton. Corks are inserted on the head-rope at intervals of 14 to 15 inches and lead, usually at the rate of a pound per *gwrhyd*, are inserted on the lead-line. In the past, stones tied in old socks or pieces of rag were used on the lead line. After setting the net on the head-rope, the lead-line is measured against it, so that the latter is 2 feet to 3 feet longer than the head-line. In setting the central *rhwyd got,* however, the head-line should be 1 foot longer than the foot-line, but for the other nets the converse is true. In *rhwyd nesa'r got,* the foot-rope is 2 feet longer than the head-rope; in the *rhwyd ganol* it is 2½ feet longer and in the *rhwyd fastwn,* the difference in length is 3 feet.

Fishing on the Teifi takes place on the ebb tide, and each team proceeds to its allotted station in the river. The sailing down river from St Dogmaels is described as *mynd i'r sân* ('going to the seine'). The teams wait on the shore near their allotted station until the tide is considered right for the first cast or *ergyd*. With the net folded at the bottom of the boat, fishing begins by casting the net as the boat is steered out to the middle of the river. Casting the net, regarded as the prerogative of the owner of the boat, is always carried out to starboard and never to the port side or over the stern. The net is paid out to a half-moon shape from the shore and the Fisheries Regulations state that 'One end of the head-rope of a draft or seine net shall be shot or paid out from a boat, which shall start from such a shore or bank, and shall return thereto without pause or delay, and the net shall thereupon forthwith, be drawn into and landed on the shore or bank on which the head-rope is being held'.[15] One member of the team, usually the least experienced, acts as a shoreman and holds the head-rope on the bank; a second, usually a man of some experience, stands in the bows of the boat and is armed with a long pole to check on the depth of water. When rowing boats were used, this process of checking depth demanded considerable skill as it was important that the insertion of the pole in the water did not retard the movement of the boat. The other two members of the team were concerned with steering the boat and paying out the net, the owner being concerned with the head-rope, his assistant with the lead-line. After the whole net had been paid out, the boat is steered towards the shore; it is then anchored and the three fishermen and the shoreman haul the net in. If a salmon or sewin is caught in the net it is knocked on the head with a wooden club or *cnocer*. It is important, say the older fishermen of St Dogmaels, that the *cnocer* should remain in the boat until the team are quite sure that there is a fish in the net. Until approximately two-thirds of a seine net is hauled in, the two pairs of fishermen haul in on the head-rope only, but the final third, two fishermen haul on the lead-line, while the other two haul the head-line.

A single *ergyd* takes approximately fifteen minutes and hauling in the net takes a further fifteen minutes, but very often it is essential that fishermen leave their homes about two hours before they start fishing. This is to make sure that there is enough water in the river for them to reach the fishing position. At neap tides, however, half an hour is usually enough for travelling from the village to the fishing grounds. 'In the past', says one of the St Dogmaels fishermen,[16] 'a mark was made to denote the beginning of a fishing session at *Y Rhyd*, as so many boats waited for an *ergyd* at that spot. The first boat had to begin fishing as soon as the condition of the tide was right, so that all boats would have an opportunity of fishing. A pole was placed in the water and as soon as that was clear of the water fishing could begin. One of the fishermen would have to go up the river bank and as soon as the first boat had completed its *ergyd*, he would make a signal for the second boat go begin its trawl . . . To get a second trawl, it was customary for the boat to move 200 yards upstream from the beginning of the first trawl.' Fishing continues for a continuous period of five or six hours and on returning to the village the first task is to have food, before the catch is weighed and packed ready for market. The fishermen are paid a piece-rate wage and the share of each team is divided into five equal parts – four for the fishermen and 'one for the boat'. The fifth share is designed to pay the expenses of the boat owner, who is in most cases a member of the fishing team. Payment is usually made on Saturday morning when a member of each team goes to see the fish dealer and returns with the money to the Teifi Inn, where the money is shared out. In the past a daily auction was held at the Netpool, three local dealers competing for the fish.

On Saturdays too, the nets that would have been kept in the boat since the preceding Monday are brought ashore to the Netpool and spread out on the net racks for cleaning and repair. In the past when cotton nets were in common use, the nets were boiled in oak bark liquor for preservation at the Netpool. This was done monthly on a Saturday, the nets being left in the boilers for twelve hours at a time.

In 1962, a religious ceremony of blessing the fishing season was revived at St Dogmaels and on 1 March of that year the service was held at *Carreg y Fendith* ('the Blessing Stone') above Pwll-y-Castell, near St Dogmaels. 'It is a local tradition', says one observer,[17] 'that the Abbot of St Dogmaels Abbey used to stand on the stone blessing the river and its harvest . . . and in later times, it was from this stone too that the vessels carrying emigrants to America from Cardigan were blessed before they sailed.' Unfortunately, it is believed by the salmon fishermen of the Teifi estuary that if a parson or preacher is seen on the foreshore, that is an omen of ill-luck.[18] Many believe that the poor salmon seasons experienced since 1962 are due almost entirely to the revival of this religious ceremony and to the presence of gentlemen of the cloth at that ceremony.[19] Someone appearing on the river bank in red clothes is also regarded as a sign of bad luck, but someone dressed in white would be very welcome.

The cry of a bird, probably a Redshank, described locally as 'a Welsh parrot', that screeches the words *'Dim byd, dim byd'* (Nothing, nothing), also signifies that the particular fishing session when it appears will be without a catch, especially if that bird flies across the river and over the fishing team. In that case, the team might as well return to the village, for nothing will be caught on that particular tide.

TYWI

The method of seine netting in the Tywi estuary is somewhat similar to that of the Teifi, although on the Tywi the boats are not usually equipped with outboard motors. Fishing is not carried out in distinct pools as on the Teifi and the rules and regulations relating to order of fishing and the location of each netsman's trawl are not so highly developed. A strict rota is followed, the first boat on the river on the first day of the fishing season being the last on the second day, and so on until the end of the season. The centre of seine net fishing on the Tywi is the village of Ferryside and fishing is carried out on the ebb tide 'between an imaginary line drawn across the said river from a point at the railway crossing gates, 1,525 yards above Ferryside station to a cottage situate at Pill Glas on the Llanstephen bank of that river, and an imaginary line drawn across the said river from Wharley Point to Ginst Point and thence to the wreck known as the Craig Winnie and thence to St Ishmael's Church'.[20] The fishing grounds are therefore barely three miles in length and a trawl usually begins at a place called Marsh gate and continues on the ebb as far down river as Ty-gwyn, at a place where the estuary of the Taf meets that of the Tywi. Although in the mid-nineteenth century the Ferryside seine-netsmen occasionally ventured as far down river as Cydweli, this part of the estuary was regarded as being very poor 'for the river is nothing but a series of pools'.[21]

Today, there are nine seine nets operating in the Tywi estuary, a considerable decline from conditions in 1861, when '22 men, 20 wives and 69 children were engaged in fishing at Ferryside'.[22] Until the nineteen-thirties, it is unlikely that the number of seine net licences on the Tywi exceeded the 15 that the Second Annual Report of the Commissioners described in 1863 and the fishermen were regarded as 'the poorest section of the com-mutiny, following a trade that was of necessity seasonal'. Until 1939, many of the fishermen were employed in the anthracite coal mines in the autumn and winter months, while others were concerned with transporting sand and gravel by boat from Ferryside to builders at Carmarthen. In the nineteen-twenties, ten boats were constantly employed in this trade.[23]

In the eighteen-sixties, according to the Report of the Commissioners,[24] the 15 teams of 'long netsmen' at Ferryside were at loggerheads with the coracle men of Carmarthen, that operated within a few hundred yards of Ferryside. In 1862, the Tywi coracle men 'went in a body to Ferryside, where the long-net men mostly live, attacked the men and destroyed their nets. A complaint was made of this outrage to the magistrate but it went no further'. Undoubtedly at that time seine netting was regarded almost as an innovation and a danger to the livelihood of the coracle netsmen and those that operated stake or mud nets in the Tywi estuary. According to one witness to the Commissioners[25] seine nets had only been introduced on the Tywi about fifteen years previously, and had replaced the stake nets that had been prohibited because they were regarded as a danger to navigation. Furthermore, the authorities were attempting to curb the activities of the coracle men as the coracle nets were regarded as being 'very destructive' in the more confined portions of the river and the coracle men were regarded as 'lawless and aggressive'. Of the seine netsmen them-selves their life was described as 'very hard; they sometimes lay out a good deal of money to purchase materials for fishing, and then they catch no fish . . . They hire out to common labour; in fact necessity compels them to do it'.[26]

A team of seine net fishermen on the Tywi consists of two people only – a boatman who rows the boat into the river and a shoreman responsible for holding the head-rope on the shore. The net, which is 200 yards long, 6 yards deep and with a mesh of 1½ inches, is carefully arranged on the stern of the rowing boat, so that it unwinds easily as the boat is taken out in the stream. The boat then returns to the shore and the net is hauled in by the two fishermen. Although a 1½ inch mesh is specified by the River Board, in practice, due to the excessive weight of such a net, the mesh is 1½ inches from knot to knot at the bottom and 2 inches at the top. Nevertheless, hauling a net of even this weight is strenuous work for two fishermen and rarely is the full length of 200 yards used. Usually, the nets are limited to a 100 yards or so. An attempt was made in 1931 to introduce three-handed fishing teams on the Tywi, but the experiment was abandoned after one season. Traditionally the proceeds of fishing are divided into three shares – one each for the fishermen and one for the boat.

TAF WADE NETS

The Taf is a river of approximately 32 miles long that runs into Carmarthen Bay near the village of Laugharne. The estuary of the river is too shallow for the use of seine net and boat, but the

Wade net fishermen on the River Taf

so-called wade-net, used by a man actually wading in the water, is used. His companion stays on the river bank and acts as a shoreman.[27] In the mid-nineteenth century, nine or ten wade nets were in constant use,[28] but today this simple form of seine net is used by one licensee only. The net is fixed to two poles and although two men are required to operate it today, in the past it was used single handed. One pole was fixed to the shore; the fisherman then waded into the stream, holding the other pole and paying the net out in a semicircle. He then returned to the shore, and with a pole in either hand, the net was hauled in.

Today, the fishing area extends from the bar of the river, 'between an imaginary line drawn straight across from Whaley Point to Ginst Point and an imaginary line drawn straight across the river in a south-westerly direction from the old Lime Kilns'.[29] The principal pools fished by the licensee and his son are Ferry House Pool, Cover Cliff Pool and Whaley Pool. In addition to fishing for salmon in the Taf itself, the wade net is used for catching flat-fish from Pendine beach also, but a licence is not required for doing this.

The wade net, operated just by two part-time fishermen, measures 30 yards long, 2 yards deep and is a plain wall of netting with a mesh of 2 inches from knot to knot. Attached to the two ends of the net are two upright poles, each 30 inches long, for holding the net in place while fishing, and also to help in hauling in the net. One fisherman enters the water on the ebb, passing the rope loop attached to the pole over his shoulders and holding the pole in both hands in a vertical position, against his chest. Meanwhile, the other fisherman stands on the shore holding the other pole in a similar manner. They gradually make their way upstream, the shoreman walking as parallel as he can to his companion. As soon as a strike is felt, the fisherman, who is often up to his neck in water, makes for the shore as rapidly as possible, and the net is hauled in by both fishermen in the same manner as a seine net. The wade net, which is mainly used at night, is not an efficient piece of equipment, for not only is the depth of water a limiting factor, but the length of net a couple of fishermen can handle whilst wading against a strong current is very limited. During the 1969 fishing season, for example, when, due to the temperature of the water, fishing could not begin until mid-June, a total of only 10 salmon and 5 sewin was caught in the Taf estuary. The cost of licence for a wade net is £4.63.

Although today river wade nets are only found on the Taf, in the nineteenth century they had a much wider distribution. They were known, for example, in Milford Haven, the Dyfi estuary and along the south coast of the Llŷn peninsula, the north Wales name for the net being *rhwyd troed* (foot-net), while in west Wales it was known as *rhwyd wad.* One or two wade nets are still used for sea fishing off some south Pembrokeshire beaches at the present time.

NEVERN AND CLEDDAU

The Nevern is a short stream, some thirteen miles long, running from the Preseli hills to the sea at Newport, Pembrokeshire. Although no commercial salmon fishing has been practised in the estuary in recent years, the present regulations allow the use of draft nets 'between the iron bridge at Newport and an imaginary line drawn straight across the said river from Dinas Head to Penybat at the estuary of Nevern'.[30]

Seine netting was stopped in 1958, despite the fact that the Nevern is regarded as a good salmon river. According to witnesses to the Commissioners appointed to inquire into salmon fisheries in 1861, seine nets as well as coracle nets and weirs were utilised on the Nevern. At Newport, said one witness, four seine nets were in regular use 'and in some seasons', he stated, 'when there has been a great take of fish, I recollect a great many coming from St Dogmaels near Cardigan. I have seen as many as five extra seines come on account of the fishing being so good'.[31] Nevertheless, during the early eighteen-sixties, there was a sharp decline in the use of seine nets in the Nevern, due to overfishing in weirs and 'from the evils resulting from the spear and small meshed nets'.[32] The result was that the number of salmon and sewin caught in the Nevern 'can scarce pay the expenses of a single draught net'. By 1870, however, four draft nets were in regular use in the estuary, while during the first quarter of the present century, three or four teams were engaged in seine netting. According to one informant[33] there were 17 fishermen at Newport in 1921. One boat was manned by a team of five and used a net 200 yards long, while the other two boats were each manned by teams of six and they used seine nets 320 yards long. As on the Teifi, a Nevern seine net was made up of a number of nets fixed together with twine. Each member of a fishing team was expected to provide a net 50 yards long and each one was carefully tied to the next (a process known as *murio'r rhwydi*), to provide one long net, each section being expected to last for three fishing seasons. Nineteen-foot rowing boats built by David Williams of Aberystwyth were the usual boats used on the Nevern before the last war and in fishing the boat owner or 'captain' was employed on the rudder, two were engaged in rowing and two were concerned with paying out the net on the starboard side in the centre of the boat. The other member of a six-man team was engaged as a shoreman. A fishing session usually began three hours after high tide when a particular stone *(Y Garreg fach)* appeared on the shore, and when one of the stone steps on the quayside was also visible at the same time. Usually, only the morning tide was fished, as night tides were considered poor. Fishing was carried out in a number of stations in the estuary, the principal ones being *Y Llygad* (the eye) or *Gene's Afon* (river mouth), *Y Dor* and *Tyn Segur*. The first boat on the first day of the fishing season fished *Y Llygad* while on the second day it went to *Y Dor* and on the third to *Tyn Segur*. A strict rotation was followed throughout the season. On incoming tides, two other fishing stations, *Benet* and *Pen Ucha'r traeth*, were also fished, and whichever boat was fishing at *Y Dor* on the ebb could proceed to *Pen Ucha'r traeth* to fish on the flood tide. The team fishing *Y Llygad* on the ebb proceeded to *Benet* on the flood. *Y Llygad* and *Pen Ucha'r traeth* were considered the most fruitful fishing stations, but only two trawls at the most could be completed on any one flood tide, a single trawl usually taking about 45 minutes.

In the nineteen-twenties the cost of the nets and licence

(£2 per 50 yards in 1921) was shared between all members of a fishing team, but the boat was usually the property of the captain of the team. The proceeds were divided equally between all the members, the salmon being sold to a Cardigan merchant.

Occasionally the weather would be unsuitable for seine netting in the estuary of the Nevern and in the sea beyond, and it was customary then for the fishermen to undo the separate sections of the net, to make a series of short 50 yard wade nets. A wade net was known locally as *rhwyd wad*, and it could be used at any time by any of the fishing teams, with the exception of the team that was allocated to *Y Llygad* on a particular day. That particular team was not allowed to go upstream with a wade net and if the weather should be too rough to use a seine at *Y Llygad*, the team allocated to that station could not fish that particular tide.

The Gwaun, a river ten miles long that enters the sea at Fishguard, was never an important salmon river, although in 1862[34] 'two or more seines were worked just outside the mouth of the river, but only one is now employed'. By the end of the nineteenth century, there was no netting carried out in the Fishguard area.

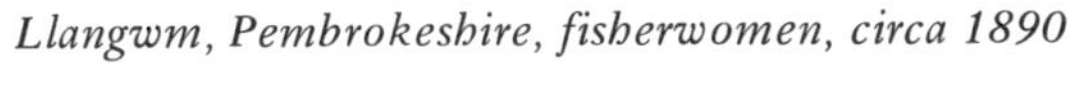

Llangwm, Pembrokeshire, fisherwomen, circa 1890

Seine netting until 1971 was carried out on the Eastern and Western Cleddau in Pembrokeshire, but as a result of a public enquiry seine netting by the two teams of four men was declared illegal. During the nineteenth century, seine netting as well as coracle fishing and compass netting was widely practised on the two Cleddaus, for, says the Second Report of the Commissioners,[35] 'these two rivers must have wonderfully productive powers to have withstood the heavy fishing to which they have been subjected for many years. Their common estuary was fished by 60 or 70 boats, using some draught nets, but principally stop nets'. In 1887 the 'Great Net' which had been abolished on the Teifi as being illegal was brought to Milford Haven and set up by the seine netsmen there. During its first season of operation, 2,000 salmon and sewin were caught in this net.[36] It only operated for a few seasons. The seine nets operating in the estuaries numbered no more than five or six throughout the nineteenth and early twentieth centuries, the netsmen being mainly from the villages of Hook and Llangwm. Throughout the period the seine nets were far less important than the stop nets, that still operate on the Western and Eastern Cleddau. In 1969, for example, seine nets caught only five salmon as compared with 72 caught by the compass netsmen.

DYFI AND DYSYNNI

The Dyfi, a river 35 miles long, is renowned for its salmon and sea trout and was described by one author[37] as 'nearly perfect as a salmon river with its abundance of rapid streams, deep pools and excellent spawning grounds'. In the mid-nineteenth century it had coracle nets and weirs as well as draft nets and despite the fact that it suffered from a certain amount of pollution from the lead mines of Montgomeryshire in the eighteen-fifties and sixties, it very rapidly regained its pre-eminence as one of the finest salmon rivers in Wales, as lead mining declined during the last quarter of the nineteenth century. In 1830, for example, the Dyfi supported six teams of three netsmen; by 1860 only two teams of part-time fishermen were at work on the Dyfi. By 1880 the river had recovered from pollution and a dozen teams were fully occupied in salmon fishing in the estuary. All the commercial salmon fishing during the present century has been carried out by seine netsmen. These are based on two stations: at Aberdyfi on the north bank of the river near the mouth, and at Glandyfi some five miles upstream where the river is considerably narrower. In 1970, Aberdyfi had six teams of seine netsmen while Glandyfi had three, each team consisting of two members only.

The Dysynni is a short river, 14 miles long, flowing into the sea at Tywyn. It has never been regarded as a particularly productive river as far as salmon and sea trout are concerned, but according to one mid-nineteenth-century report[38] it was remarkable for the large size of its fish. 'It passes through the lake called Tal-y-llyn', said the report, with the result that it was a very slow-running river; 'the salmon in it are consequently larger and earlier; 17lb and 18lb fish are not very uncommon and they reach the weight of 32lb and occasionally a fish is taken in May. The white trout or sewin are also stated to be remarkably large, the early runs being from 4lb to 10lb each'. Two fishing weirs at Ynysmaengwyn and at Peniarth were in use until about 1861, while in that year 'four or five draught nets' were in use. Nevertheless, due to the slow-running nature of the Dysynni and the fact that it flows into the sea through a narrow, shallow bottleneck channel, some restriction had to be placed on seine netting in the river in the late nineteenth century. 'The salmon usually hang about at the mouth until a flood comes', said one observer,[39] 'and during this period of waiting, they used to be swept up by the netters until 1887 when a bye-law was made prohibiting any netting within three hundred yards of the mouth of the river'. It is doubtful whether at any time since 1887 more than two teams of netsmen have been employed in the Dysynni estuary. In 1970 two seine nets for the Dysynni were issued, but they were only very sporadically used by two pairs of part-time netsmen.

Unlike the seine net fishermen of the Teifi those of the Dyfi, and indeed all the 31 netsmen licensed by the Gwynedd River Authority, are prohibited by law from using motor boats, and

Seine netting on the Dyfi at Glandyfi, 1971

12 foot or 14 foot rowing boats of shallow draft are universally used throughout the area. The number of seine netsmen has declined considerably in recent years, for in the nineteen-twenties there were at least twenty full-time fishermen at Aberdyfi fishing on day and night tides for continuous periods of eighteen hours or more. A shift system was worked with perhaps one team working from midday to 10.00pm to be replaced by another team working from 1.00am to 11.00am. A team would work the night tides for a week and the day tides for the following week, each team consisting of three people. Two would be employed in a boat and another as a shoreman *(tantiwr)*. Today most of the fishermen are part-timers, each team consisting of two persons only — a boatman and shoreman.

Unlike the Teifi netsmen, there are no set rules on the Dyfi regarding the location of the fishing teams in the river, but there is a great deal of co-operation between members of fishing teams who help one another with hauling the nets in and re-arranging the net on the transom of the boat. The first boat in a pool takes the first *ergyd* with a net that measures 150 yards long, 4 yards deep with a 2 inch mesh. In the past nets of 200 yards in length were allowed, and before each *ergyd* it was customary for the fishermen to spit on the net as a token of good luck. In practice, and particularly in the narrower stretches of the river around Glandyfi, the full legal length of net is rarely used and the seines are often no more than 100 yards long. With the steeply sloping banks, the head- and foot-ropes may extend as much as 30 feet beyond the mesh of the net, although the full length may not be required when the tide is well in and the fishermen can haul the net while close to the water. When hemp and cotton nets were used, Dyfi seines were equipped with two head-ropes *(tannau uchaf,* sing. *tant uchaf)* and two foot-ropes *(tannau isaf,* sing. *tant isaf)* to prevent kinking while arranging the net on the boat transom.

As on the Teifi, salmon fishing is carried out in a number of specific pools *(llynia')*, the principal ones being Y Garreg, Ergyd y Station and Pont y junction above Glandyfi; Ergyd Tom Pugh, Domen Las, Ergyd Cadi and Ffôs Derfyn in the narrow meandering section below Glandyfi. Downstream in the wider estuarine waters between Aberdyfi and Glandyfi on the north bank of the river are Bwdlin, Frongoch, Tre Fry, Plas y Bishops and Aberafon. While Aberdyfi fishermen rarely venture beyond Pwll Bwdlin, those of Glandyfi, after a short time at Ergyd y Station, often drift downstream on the ebbing tide to the estuary, returning to Glandyfi as far as the railway bridge on the flood.

Fishing is carried out usually for continuous periods from half flood to half ebb and at Aberdyfi the flood tide lasts for seven hours and the ebb tide for six. At Glandyfi the narrow river means that tides are some 9 feet higher than in the estuary, while the flood lasts for about two hours as compared with six in the estuary. The pools in the estuary can therefore be used for considerably longer periods than those in the Glandyfi section of the river. When the tide is strong, fishing is difficult, for the foot-rope becomes too high in the water and salmon can easily slip beneath the net. In warm weather the lower pools are regarded as the best for fishing.

The fishing season extends officially from 1 March to 31 August, but in practice few venture on the river before April. One man is concerned with rowing the boat into the stream, another with holding the head- and foot-ropes on the shore. The net, which has been carefully arranged on a wooden plank or sheet of tin on the boat, unwinds as the boat is rowed into the stream. When the river is flooding, the fisherman steers the boat downstream, turning to starboard when about half the net has been paid out. He then makes his way back to the shore and lands within about five yards of the *tantiwr*. At night it is the shoreman's constant whistling that acts as a guide to the boatman as to where he should land. The boat is anchored and the net, forming a semi-circle in the water, is hauled in. Members of a second team of netsmen help with the hauling of the net. While two are concerned with hauling the head-rope with the cord or plastic floats, the other two are concerned with hauling the foot-rope which carries lead weights. The former pair stand, the latter kneel or crouch, pulling in the foot-rope as close as possible to the bed of the river. Since the two pairs of fishermen haul in the net at different heights, a distinct bag is formed in the net, especially as the head-rope is pulled in at a slightly quicker rate than the foot-rope. Any salmon or sewin caught in the mesh finds it difficult to escape as the net is hauled in. The enmeshed fish is killed with a length of lead piping or a wooden knocker *(huwcyn),* but on the Dyfi it is considered unlucky for a fisherman to fetch the knocker from the boat before the fish is seen. If he does so, then he must not use the knocker for the remainder of that fishing session.

Before the second team begins to fish – for on the Dyfi every team must take its turn *(ergyd am ergyd)* – the net has to be rearranged carefully on the boat's stern in such a way that it unwinds smoothly and correctly on the next trawl. When each fishing team on the Dyfi consisted of three members only, the most experienced were allowed to rearrange the net on the boat transom, a process described as *mynd i ben y bwrdd* (going to the top of the table). The fishermen, in rearranging a net, stand on the boat platform, and while one is concerned with arranging the head-rope, the other is arranging the foot-rope. Two members of another team are concerned with disentangling the net in the water behind the boat. The net is therefore arranged in two distinct heaps on the stern of the boat. It is then the turn of the second team to fish, while members of the first team take a rest before assisting to haul the net in. Each *ergyd* takes about fifteen minutes.

The process of fishing on the ebb is slightly different, for the boat is not taken downstream as on the flood, but upstream. When about half the net has been paid out, the boat is turned downstream and makes its way back to the shore, landing below the *tantiwr*. The pool near the Dovey Junction railway bridge is never operated except on the ebbing tide but all the others may be fished on both the flood and ebb.

Unlike the other Welsh seine netsmen, the Dyfi fishermen believe the night tides are better for fishing than day tides. Since they operate both in the day and night, the fishermen spend the time between tides in a hut on the river bank. These huts are equipped with bunk beds and cooking equipment and are in constant use except at week-ends, when the river is closed to fishermen.

The fish are either sold locally or sent to Birmingham, the proceeds being divided into three shares – two for the fishermen and one for the boat. When six nets were fully employed on the Dyfi, one half of the proceeds went to the nets and boat and the other half was shared equally between the three members of each team.

GLASLYN

At Porthmadog, two teams, each consisting of four part-time netsmen, operate in the River Glaslyn between the road bridge and the sea. The *Second Report on Salmon Fisheries* of 1863[40] noted that the Glaslyn was one of the few rivers in north Wales where 'there were no dams to obstruct the fish and subject them to overcapture and no stake nets to oppose their entrance to the river, and in the Glaslyn . . . which has not been abused by the erection of impediments, we still find a fair stock of fish'. At that time seine nets were widely used for the capture of salmon and four boats were in constant use in the river estuary, although 'the nets of three of them however having an illegal mesh were seized by the police in the spring of 1862 and condemned'. Earlier in the nineteenth century, the Glaslyn supported eight teams of netsmen, each team consisting of four members and the crews of the eight boats 'lived entirely upon the produce of the salmon fisheries'.[41] 'In those days', said a witness to the Commissioners in 1861,[42] 'they took a cartload of fish every tide, from the beginning of June till the end of August and they sold them at 4d per lb.'

Three full-time teams of seine netsmen still operated until 1939 while another team of four part-time fishermen occasionally fished the estuary. The nets used measure 100 yards to 120 yards in length and 20 feet deep and they are known locally as *rhwydi sin.* Each net is connected by a rope at either end to wooden stakes, approximately 4 feet long, that facilitate hauling. The rope connecting the net to the shore is known as *rhaff lan* and measures up to 50 yards long; while the rope at the other end, known as *rhaff cwch*, may be as long as 100 yards. The stakes are known respectively as *polyn lan* and *polyn cwch*. The long lengths of rope attached to the net is particularly useful in the Glaslyn estuary, where the banks on the ebb tide slope gently and where it is important to keep the main body of the net in deeper water, well away from the river bank.

The two teams of fishermen operate in a number of pools – Twll Bont, Banc Submarine, Banc Cei Balast, Trwyn Cae'r Ogo, Cwt Powdwr and Pwll Glanmor. They fish from half tide to half tide. When Porthmadog supported 20 full-time mussel fishermen in the nineteen-twenties, boxes of mussels were stacked at low tide in the river to be cleaned by the in-coming tide. As soon as these boxes of mussels appeared above the surface of the water, this was an indication that seine netting could begin with the ebbing tide. Today a heap of stones near the quay acts as an indicator of the time when fishing can begin. One member of the team acts as a shoreman, while the other three are occupied in the boat: two on the oars, known as the bow oar and the aft oar, and 'the captain' paying out the net over the stern of the boat. The fishing season for salmon on the Glaslyn extends from early May to the end of August, the best months for salmon being July and August. Many of the old fishermen believed that the turning tide *(croen llanw)* was important for fishing, for it was said that salmon always entered the river before the turn of the tide, when the water was at its lowest ebb. Fishing teams take it in turns to cast the net if they are in close proximity to one another and there is usually co-operation between the members of each team in hauling in the net.

As in other districts, the seine netsmen of Porthmadog had their beliefs and customs. It was customary for each member of the team to carry a knocker *(pren lladd)* but it was considered unlucky to carry it in the boat. Many of the old fishermen wore clogs that were removed and used for killing a salmon instead of the usual knocker. A fisherman bringing a sack to the river in order to place the fish in it for carrying was considered to be bringing ill-luck to his fellow fishermen. Warm, showery weather with periodic sunny periods was always considered the best for salmon fishing for 'the smell of rain attracted the salmon to enter the estuary at Porthmadog'.[43]

The proceeds of fishing was divided equally between the four members of a team, the price of the licence being taken at the beginning of a season. If, for some reason or other, the owner of a boat and net was unable to fish at a particular tide, he was nevertheless given his share by the fishermen that had been fishing that day.

The same method of fishing is used at the mouth of the Dwyfor where a single team of four part-time fishermen operate occasionally in a single pool at the mouth of the river.

In the Llŷn peninsula, part-timers also operate on the sea shore around Aberdaron and Nefyn, but the catches of salmon in that area are very low indeed. The coastal nets operated by three part-time teams of four netsmen each are known as *Rhwydi mawr* (large nets) and are used mainly at night. In the past no boat was used and the net was taken out into deep water by a netsman who actually entered the water. Again the nets were equipped with long connecting ropes and two upright poles.

SEIONT

Between the mouth of the Seiont and Abermenai Point on the Caernarvonshire shore of the Menai Straits, four teams of part-time seine netsmen are concerned with salmon fishing. Each team consists of four members, two being necessary to row the 14 foot boats in the strong currents of the straits. Another member of each team is concerned with paying out the net from the stern of the boat, while the fourth acts as a shoreman *(dyn rhaff lan).* In the nineteen-thirties, seven teams of four full-time fishermen based on Caernarvon dock operated in the district. Fishing is carried out in four stations *(patchis)* near the south shore of the straits. They are Belan, Ty Calch, Porth Llidiog and Glasdwr. Fishing can be carried out continuously for a session of four hours in the pools, with the exception of Belan, where the state of the tide makes it possible to fish continuously for six hours, if the tide is under 16 feet in height. Should it be more than 16 feet, however, the Belan pool cannot be used.

The rules and customs relating to the allocation of pools in the western section of the Menai Straits are complicated. The first boat to venture from Caernarvon to the fishing grounds at the beginning of the season, usually in mid-April, is designated 'Boat One' for the remainder of that season, and has to fish continuously at Ty Calch. It can move to Belan pool, however, if the team so wishes, provided that Belan is unoccupied after three hours of ebb. The second boat to venture out at the beginning of the season is designated 'Boat Two' and it too has to go to Ty Calch, but is free to move on to Porth Llidiog if that patch is not occupied. Boats Three and Four are assigned to Glasdwr, where tides are usually over 16 feet in height, but they too are allowed to move; in this case to the shallower Belan, if Belan be unoccupied after three hours of ebb. If any boat assigned to any station passes its own patch for Belan, but finds on arrival that the swell is too great to fish there, that team is not allowed back to its own station until high tide. At Belan, the tides are usually very strong and when using a seine net, the oarsmen have to row with rapid strokes or the net will overturn the boat. When seven boats fished from Caernarvon, the allocation was 3 boats to Ty Calch, 2 to Glasdwr and 2 to Belan, with Porth Llidiog being unoccupied except as an alternative to the other stations. At Belan and Porth Llidiog, the fishermen can fish continuously, if the swell is not too great, at almost every state of the tide.

The seine net, known locally as a *rhwyd dynnu,* measures 150 yards long and is 12 to 14 feet deep. The mesh measures 2 inches from knot to knot. In some of the pools, notably Belan, the whole length of net is not used and perhaps a half of its length is not paid out from the boat on any one shot. When the current is strong, pieces of slate (slipers) are attached to the foot-rope *(tant isaf)* of the net, in addition to the lead weights. Attached to both the foot-rope and head-rope *(tant uchaf* or *tant cyrcs)* at either end of the net are two lengths of rope, each 16 yards or more in length, which are firmly attached to poles.

The one pole 3 feet long is known as *y Polyn lan* and is held by the shoreman *(dyn rhaff lan)* while the boat makes its way into the fishing pools. In a shallow, gently shelving pool, such as Glasdwr, the full length of rope connecting the pole to the net is required, but when the shore shelves steeply, the ropes are wound around the pole, so that the actual end of the net is close to the shore. The other pole – *polyn allan,* measuring 4 feet or 5 feet in length – is carried in the boat and is only used when the net is being hauled in by the team of netsmen.

A heap of stones *(tocia)*, surmounted by a wooden peg, marks each fishing station, and the peg must be visible before fishing can begin. The stone heaps must never be disturbed, and if a stone should be caught up in the net at any one time, it must not be placed on top of the *tocyn,* but on its side, so as not to alter the height of that heap of stones. If the top of a heap of stones and its peg are visible, and there is no team fishing at that station, any other team may fish at that station and has priority for that particular tide.

Two heaps of stones, 250 yards apart, mark the position of the beginning and end of each trawl. The boat leaves one heap – *tocyn trai,* with the shoreman standing nearby – and makes its way out to the centre of the pool. The shoreman holding the *polyn lan* walks towards a second heap of stones, *y tocyn llanw,* when the boat is due to land again. The shoreman as he walks issues orders to the men in the boat and he has far more authority over the team at each trawl than shoremen on any other Welsh river. The boat is allowed to drift out with the tide for three minutes before it makes its way back to the shore.

The net is then hauled in, with the members of the team in a standing position hauling in the head-rope, the other two in a kneeling position hauling in the foot-rope. If there are signs that a salmon has been caught in the net, a member of the team fetches the wooden knocker *(pren lladd)* and kills the salmon by knocking it firmly on the head between the eyes. One person is in charge of the knocker at each tide, but should a fishing session be unsuccessful, that person must hand over the knocker to another member of the team for the following tide, for the unsuccessful keeper of the knocker is regarded as a 'Jonah' and brings no luck to the fishermen.

Each *ergyd* takes some eight minutes, and after every one the net has to be rearranged carefully on the boat transom; one heap for the foot-rope, the other for the head-rope. In rearranging, a process that takes at least 8 minutes, the net has to be cleaned by turning it over frequently to get rid of the kelp *(brwal),* twigs and other material caught up in it. The proceeds of fishing are divided into 6 equal parts for each team – one share for each fisherman, one for the boat and one for the net.

OGWEN

In the estuary of the Ogwen, a river that enters the sea at Port Penrhyn near Bangor, and along the coast to the Aber district, two teams, each consisting of two men, use seine nets. Flounder fishing occupies the activity of the two teams during the spring months, but as soon as salmon are seen in the river, seine netting begins, usually around 1 May. Before 1939 when three teams, each consisting of three members, fished regularly in the Ogwen, it was usual for the three nets, each 150 yards long, to be attached to one another and stretched right across the mouth of the river to form a weir. This was especially effective when the tide was running strongly.

Fishing takes place in distinct stations – Yr Elba, Y Gwryddyn, Y Banc and Y Bonyn Bach – and usually after carefully arranging the net on the gunwale, the rope at its end is attached to a heap of stones *(tocia)* or a tree stump *(y bonyn bach)* thus allowing the two fishermen to row out into the river. The first boat out during the fishing season fished Elba on the first tide and would proceed to Gwryddyn on the second. The order of fishing is strictly adhered to during the whole season, the teams only being allowed to fish for one tide at any one station. The proceeds are divided into equal shares, one each to the fishermen, one to the boat and one to the net.

CONWAY

The Conway, famous for the large size of its sea trout, has six seine net teams that operate between Caerhun and the sea.[44] In 1970 the number of teams was eleven. Each team consists of four members, but more often than not only two fishermen, one in the boat, the other on the shore, are employed during a fishing session. All the fishermen are full-time workers, being employed in mussel fishing in the estuary in the winter months, in-shore fishing and sparling fishing in the spring and seine netting for salmon in the summer. Two of the boats usually fish at Tal-y-cafn, two up-river at Caerhun and the third pair at the bar of the river. The last station is said to be particularly good in hot weather, when salmon tend not to venture upstream. Tradition has dictated that each team fishes its own station and never ventures to another's preserve. Thus, for example, William O. Jones, who took over a netting licence from an elderly fisherman in 1957, always fishes at Caerhun, as the previous licence holder did. For sparling fishing in April, his fishing station is near Tal-y-cafn bridge, while in winter he is concerned with mussel gathering in the estuary.

The salmon seine net measures 100 yards long and is equipped with cork and plastic floats, each 24 inches apart, and lead weights, each approximately 36 inches from the next. The net is set in by the third, as is common in most seine nets, and in order to prepare a net of 100 yards in length, it is usual to order one of 125 yards in length; the nets are usually obtained from a Scottish manufacturer. Before casting a net at the beginning of each fishing session, it is customary on the Conway to throw water on the net so that it sinks better. The net is carefully stowed on the boat transom, one member of the team responsible for coiling the lead line, the other for the head-rope; the net is grasped in 18 inch lengths. The net is therefore arranged in two distinct heaps and is connected by a rope *(y pen lan)*, approximately 4 yards long, which is held by the shoreman *(dyn lan)*. Another rope *(rhaff hir)* 100 yards long is attached to the seat of the boat. The junior member of the team is responsible for rowing the boat into the stream, while the senior member acts as a shoreman. Fishing usually continues for two or three hours, for at the turn of the tide the currents are too strong to allow the operation of a seine net. Two tides are usually fished, a single trawl taking approximately fifteen minutes. If two teams operate at the same fishing station, they take it in turns to cast the net, the second team not being allowed to fish before the other has completed its trawl. Conway salmon are usually sold locally after each tide and the proceeds of sale are divided equally between members of the fishing team.

The senior member of each team usually carries a hardwood knocker *(pren lladd)* in his pocket and no superstitions associated with the possession of that particular piece of equipment are associated with Conway seine netsmen. Nevertheless it was believed in the past that the possession of a piece of coal or a bone by the fishermen always brought good luck to the team. As far as another fish, the sparling, is concerned, however, there are many superstitions. In recent years, the sparling *(brwyniaid – singular brwyniad)* is caught on the Conway only, though in the mid-nineteenth century they were caught on other rivers in north Wales, such as the Dee and Clwyd.[45] Local tradition states, however, that sparlings occur on no river in the world but the Conway, and even on that river they do not appear until all the snow on the mountain peaks has disappeared. The fish is connected by legend with Saint Brigid or Sant Ffraid, who sailed from Ireland on a piece of detached ground and landed on the foreshore at Glan Conwy, where the ground became part of the coast on which she built a church. Legend says too that when there was great famine in north Wales, Saint Brigid grasped a bundle of reeds *(brwyn),* and cast them into the river. These turned into *brwyniaid,* which provided food for the starving population.

Sparlings are caught with a seine net 100 yards long, 2 yards deep and with a mesh of ¾ inch from knot to knot. Fishing takes place at Tal-y-cafn in late March or early April and although today only two pairs of fishermen are concerned with sparling fishing, before 1939 it was usual to see eleven boats at work below the bridge. In the past, the fish were sold on the quay at Conway, although it was customary to present the first catch of the season to Lord Aberconway. No licence is necessary to fish for sparling and the netsmen use the same 10 foot boats that are used to fish for salmon.

DEE

The Dee has always been one of the most important salmon rivers in Wales, and the river still supports 31 teams of seine netsmen in addition to 4 teams of trammel netsmen at Flint. The draft or seine netsmen are based on two centres – Chester and Connah's Quay. Legally, nets up to 200 yards in length, 5 yards deep and with a 2 inch mesh may be used, although in practice, the nets used are 170 yards long, 4 feet 6 inches deep with a mesh that varies from 2 inches in the centre to 4 inches at the sides. This is to lighten the weight of the net, for although a licensee may endorse up to four others on his licence, in practice the net is operated by two fishermen only. In the Connah's Quay district, where the river is considerably wider than in the Chester area, three- or four-man teams are usual. Again, as on the Conway, 16 foot or 18 foot rowing boats are used for seine netting during a season that extends from 1 March to 31 August. The river is a fairly late one, for although a run of salmon may be experienced in March and April, most of the fish are caught in July and August. In 1970, for example, the following quantities were caught:

MONTH	CHESTER DRAFT NETS Number	Weight (lbs)	CONNAH'S QUAY DRAFT NETS Number	Weight (lbs)
March	8	101¼	6	81
April	9	115	9	87¾
May	31	318½	8	72¾
June	97	868¼	53	376½
July	535	3594½	381	2457¾
August	266	1876½	116	777¾
Total	946	6874	573	3853½

There has been a decline, both in the number of fish caught on the Dee and in the number of seine nets operating in the river. In the eighteen-forties, the number of seine nets was approximately 9 but by 1860 they had increased to 40.[46]

By 1866 47 teams were operating in the river, and each took on average 17 fish daily throughout the month of August, a total of 15,000 fish. Between 1870 and 1881, an average of 52 seine netsmen, paying a licence fee of £5 per annum each, operated from Connah's Quay and Chester. The eighteen-eighties saw a rapid increase in the number of seine netsmen. In 1882-3, 63 were licensed; in 1884-5 and in 1885 the number had reached the all time high figure of 96 boats. By 1888, however, navigation improvements on the river between Connah's Quay and Chester prevented many from working and in that year only 59 boats were licensed.

It was estimated that in the eighteen-sixties, 280 men were regularly employed in fishing with seine nets and trammel nets in the Dee below Chester, while in addition there were '30 or 40 men called scowbankers, who otherwise employed during the week, pay the estuary a visit on Saturday and Sunday, hire or borrow boats and fish away to the great annoyance of the regular fishermen'. In 1969, the river supported 125 seine and trammel netsmen, many of them being full-time fishermen.
The seine net catches in 1970 are shown in Table 11 *(see page 258).*

SEVERN

The four teams of long netsmen operating in the Severn in the Deershurst district below Tewkesbury have already been mentioned, but until the early nineteen-fifties, seine netting was far more widespread than it is today. For example, at Elmore, some four miles down river from Gloucester, a number of teams were operating from the hamlet of Stonebench. A pool known as 'Madam Pool' was regarded as the best long-net fishing on the Severn for 'it is never disturbed by passing river traffic; the river at this point is exceptionally narrow and the drag is not more than two hundred yards long'.[47] Each team of long netsmen consisted of two boatmen and a muntle-man (shoreman). Until 1937 long-netting was practised even further downstream at Framilode.

Above Gloucester, especially in the Deershurst, Chaceley and Apperley districts where long netsmen still operate, this type of

TABLE 11
SEINE NET CATCHES 1970

	SALMON		SEWIN	
	Number	Weight	Number	Weight
Teifi	802	7666lb	254	830lb
Tywi	189	1608lb	322	1021lb
Dau Cleddau	No catches - seine nets banned			
Taf (Wade Net)	10	74lb	6	14lb
Coastal Seines	–	–	22	46lb
Dee (Chester Drafts)	946	6874lb (including salmon & sewin)		
Dee (Connah's Quay Drafts)	573	3853¼lb "	"	" "
Dyfi, Dysynni, Mawddach, Dwyryd, Glaslyn	322	2346lb	675	2353lb
South	116	977lb	302	272lb
Seiont	512	3034lb	56	244lb
Ogwen (incl. Anglesey rivers)	148	815lb	98	381lb
Conway	136	936lb	37	159lb
Total	3754		1772	

fishing instrument is the only one allowed by law for catching salmon in the deep-water reaches between Tewkesbury and Minsterworth. Each long net used is 90 yards long and 4 yards deep and until about 1955 most of the nets were equipped with cods: 'great tapering bags in the nets like the feet of socks. It takes an experienced hand two months to knit one of these nets', says one author,[48] 'and a net will only last one season, so knitting must begin before the storms of winter have winnowed the last autumnal leaves from the oak, if the net is to be ready for the opening of the season on 2nd February . . . In the last week of January, the fishermen, for it takes a team of four to work a long net, sweep the river. A long-net is worked over a quarter of a mile of water; here the bed of the river is dragged with a heavy chain to remove all waterlogged timber and the roots of trees, which have become lodged there in the course of winter floods.'

Waters describes the method of long netting as follows: 'On 2nd February the long-net is placed in the river for the first time; it is paid out from a flat-bottomed punt, being weighted to the bottom of the river by a series of leads and buoyed to the surface by a number of corks. The line of corks runs diagonally across the surface of the river to prevent the cod from being turned inside out by the force of the current. The net is controlled from either bank by two bridles, one fastened to the bottom, the other to the top of each side of the net. The net, slanting in a crescent across the river, placed the bridles on the Apperley bank further downstream than those on the Chaceley bank across the river. Thus the Apperley side of the net has to take the greater strain; for this reason these bridles are longer and are held together by a long line known as the muntle. In this position the muntle line is hauled over a quarter of a mile down the river bank with the current, while those holding the opposite bridles keep pace on the Chaceley side of the water. In motion the swill swells, the cod becomes rigid, and there is little hope for the oncoming fish. Upwards, ever upwards, is the one instinct of the salmon ascending towards his spawning grounds. The tapering swill and narrower cod make him think he will get through this obstacle. A very strong fish sometimes breaks through a net, but apart from this he has little chance of getting away if he happens to be in a reach of the river where a long-net is in operation. In this part of the river salmon swim deeply and in the opaque water are invisible to the eye. When the long-net has been drawn over its appointed quarter of a mile, the muntle is carried across the river, where a windlass draws the net into a horseshoe against the bank. In this position the current carries the cod downstream as far as a stage, which has been built for the purpose of landing the net. This stage is called the flake and differs from an ordinary river landing-stage, for its timbers, instead of running at right angles to the stream, are parallel to the river so as to give purchase to the fishermen's feet when landing the net. The cod with its catch is pulled to the flake and up the bank, where the fish are killed with a knobbler, a fifteen-inch wooden truncheon, made whenever possible of laburnum wood on account of its exceptional hardness.

'Three or four salmon in the cod are here considered an excellent haul, though below Gloucester catches tend to be more plentiful, and one old cowman at Longney assures me that in his young days he assisted in a drag when twenty-one salmon were landed. The great enemy of long-net fishing above Gloucester is the increase of barge traffic; horse-drawn barges were content to wait while fishermen finished their drag. The motor-barge gives warning of its coming and, unless nearing the end of their drag, fishermen are expected to make way for its passage. The motor-barge pollutes the river to the detriment of salmon, and fish are often mutilated through being cut by passing propellers.

'But the greatest deterrent to long-net fishing and salmon fishing generally, along the length of tidal Severn, has been the cost of a licence. The riverman will give ungrudgingly of his skill and labour to knit a long-net, but resents having to pay £10 for the privilege of working the long-net for a short season between February and August. The countryman is not a gambler, and the riverside peasant with little or no working capital resents throwing £10 in the river for doubtful returns. This sum would keep him and his family for a month and "a bird in the hand is worth two in the bush". The short-sighted policy of high licences, despite the possible returns of a good season's fishing, has done much to diminish a time-honoured occupation in which chance plays so large a part. There is real danger that the craft of making and the skill of using the long-net, together with the use of the lave-net in the lower reaches of the river, will vanish from Severnside before the end of the century.'

SEINE NETTING IN OTHER RIVERS

The methods of seine netting adopted by fishermen in rivers outside Wales are substantially the same as those described. Apart from a small fishery for sea trout at Kessingland off the Norfolk coast, there are virtually no salmon fisheries at all on the coast between the Humber and Southampton Water.

In those areas in the west and north-west of England where seine netting is practised, there are many local variations in the size of net and the number of men operating in each fishing team, the methods being largely dictated by the regulations of the various River Authorities. In the Devon River Authority's area, for example, there were 81 seine net licences issued in 1971. 36 of these operated in the Taw and Torridge rivers during a fishing season that extends from 1 April to 31 August. Each net is 200 yards long and of a depth not exceeding 8 yards, and each one is equipped with two end poles as in the nets of north Wales.

On the River Dart 15 seine nets, each manned by a team of three operate in the estuary, principally from Stoke Gabriel on the eastern bank of the river. Seine netting is carried out by a group of men, who work part of their time as farm labourers and are also employed in cockle gathering on the sandbanks in the river and in netting mackerel, mullet, bass and other fish. The boats they use are 16 foot carvel-built rowing boats. Although each team only consists of three members, nevertheless a team operating at any one time at Stoke Gabriel has the services of a second shoreman, who assists all the teams in hauling in the net.

Appreciable quantities of both salmon and sea trout are caught in the rivers that come under the jurisdiction of the Devon River Authority as is shown in Table 12 *(see page 262).*

In the Avon and Dorset River Authorities' area, seine netting is practised in the joint estuaries of the Avon and Stour and the Frome and Piddle. Each team consists of either two or three members involving about 30 part-time fishermen and catches vary between 700 and 1,200 salmon per season.

In Cornwall seine netting is practised in the estuaries of the Tamar, Lynher, Tavy and Fowey, the number of licences in 1971 being Tavy – 3, Tamar – 14, Tavy/Lynher – 2, Tamar/Tavy – 5, Lynher – 3, and Fowey – 4.

Up to 4,000 salmon and 1,400 sea trout are regularly caught in seine nets in Cornish river estuaries during a season that extends from 1 March to 31 August. Each net according to the River Authority's regulations is limited to 200 yards in length and 10 yards in depth.

In Lancashire River Authority's area seine or draw nets are fished on the Duddon and Lune estuaries. 'They used also to be fished on the Kent and Leven estuaries, but during the 1914-1918 war, the area in which netting was prohibited was extended seawards to exclude those estuaries.'[49] The season extends from 1 April to 31 August and each seine netting team consists of three members, a shoreman and two others who row the boat. The minimum mesh allowed is 2½ inches and the nets are up to 200 yards in length. On the Duddon, two fishing teams regularly fish the estuary 'but on the Lune there are only two reaches that are worth trying with a draw net and one net fishes these'.[50]

Further north in Cumberland one seine net team fishes the Esk and 3 the Eden, in a season that extends from 1 April to 31 August.

TABLE 12

Number of salmon and sea trout caught by seine nets in Devon 1970

RIVER	LICENCES	SALMON CAUGHT	SEA TROUT CAUGHT
Avon	2	187	29
Axe	1	16	6
Dart	15	410	186
Exe	17	2087	3
Taw & Torridge	36	2040	1724
Teign	10	1946	22

Number of salmon and sea trout caught by seine nets in previous four years

RIVER	1966	1967	1968	1969
Avon	116	134	59	51
Axe	–	55	6	–
Dart	1262	1188	541	654
Exe	2398	2428	1086	1480
Taw & Torridge	2557	2038	1915	2428
Teign	1303	1539	1016	1484

For notes to Chapter 9, see pages 321-3

10 Eel Capture

The common, or freshwater, eel *(anguilla anguilla)* occurs widely throughout Western Europe, and it is found in rivers and inland waters throughout Britain. In whatever stretch of water eels are found, it is certain that they have travelled there from their breeding ground in the Sargasso Sea, where they breed at a depth of between five or six hundred fathoms in a remote area between Bermuda and the Leeward Islands. It is here, alone, that young eel larvae have been found, and it is to this area too that adult eels will return to breed and die. Immediately after hatching, the larvae in their millions begin their long journey eastwards to Western Europe, a journey that takes nearly three years. The larvae, known as *Leptocephalids,* are transparent and laterally flattened and they begin their journey of over two thousand miles at a depth of less than fifty fathoms. When the larvae are nearing the coast in the autumn of the third year they change into young eels or elvers and enter the rivers about March or April in countless numbers. The elvers are formed by a decrease in the length and depth of the larvae and their bodies become cylindrical. They are then unmistakably young eels, between 2 inches and 2½ inches in length.[1]

In the lower Severn Valley, below Gloucester, elvers are considered a delicacy and, until recently, vast quantities were caught every year as they entered the river with the Severn Bore. 'The coastline of the Bristol Channel', says Taylor,[2] 'forms a vast funnel in the path of the migrating elvers and far greater numbers enter the Severn than any other river in the British Isles. For centuries elvers have been greatly esteemed as a delicacy in Gloucester and the district bordering the tidal reaches of the lower Severn.'

Most of the elvers that enter British rivers, however, spread upstream to brooks, canals, ditches and even reach isolated ponds. In these places they feed and grow until they reach maturity, between eight and twelve years of age. Eels that are feeding and growing are known as 'yellow eels' or 'gelps' and are green or yellow in colour on their undersides. Yellow eels were often caught in the past, and a variety of instruments ranging from tongs to spears were used in their capture.

Nevertheless, most of the eels that are, and were caught in the past are silver eels[3] that change colour to silver when they are about to migrate to the sea. The migration of the silver eels takes place in the autumn, the actual time of migration varying considerably with the locality and the nature of the season. The main run is usually during the autumn floods, especially on dark nights. 'These silver eels are in splendid condition', says a Government Report;[4] 'they are on their way back to the sea and

Elver fishing at Northfield Stack, Longney

will never come back; their capture will have practically no effect on the future stock of eels in the particular river in which they are caught, for plenty of eels will reach the breeding places from elsewhere. For these reasons, as many silver eels as possible should be caught, and this is comparatively a simple matter, for, like other fishes that migrate in numbers at a certain season, they can be intercepted on their journey at selected points.'

ELVER FISHING[5]

Elver fishing during the spring seems to have been limited to the lower Severn, especially the stretch of river below Tewkesbury. Although it is difficult to estimate the exact quantity of elvers caught or the number of riverside dwellers engaged in elver fishing, it is certain that appreciable quantities are caught. During a poor season like that of 1971, when 'the elver run was again reported to be poor . . . several catches of one hundredweight were made'.[6] Unusually large catches were often made in the past. 'On 24 March, 1943, for instance, four men fishing just below Tewkesbury took three hundredweight. 1944 seems to have been an exceptional year, for on 26 March there were about fifteen hundredweight of elvers (approximately 1,680lb) in sacks on the banks of the river near Tewkesbury awaiting transport. In the same year, 1,400lb were taken at Framilode between 2 and 6 April. The first elvers to come up river at the beginning of the season are said to be heavier than those in later runs. Reliable counts have given between 90 and 100 to the ounce. A catch of a hundredweight, therefore, represents a total of something like 170,000 individuals. Since a hundred or more men may be fishing from the banks of the river between Tewkesbury and Epney on a single night, total catches must be reckoned in many millions.'[7]

Elver fishing is usually carried out on the spring tides at night, although occasionally day tides are fished. The best fishing is done just after the tide has turned to the ebb although good catches are sometimes made before the tide. The fisherman chooses his fishing position carefully, usually selecting a solid, grassy tump as near the river as possible and at a point where

An elver fisherman with elver fishing equipment

the river flows strongly. Elvers hardly ever enter slack water and they always swim against the flow of the tide. The fisherman, after selecting his fishing position before the arrival of the bore, tries the water with his net to see if there are any elvers about. 'Presently the warning cry of "tide" is heard down river. Each fisherman in turn takes up the cry and passes it on. Then, as the bore approaches, he scrambles up the bank with his gear out of harm's way. Once the bore has passed, there follows an anxious wait of about an hour until the tide turns to the ebb. This is the moment when fishing begins in earnest, and if the elvers are running well, a single fisherman may take a hundredweight or more in a matter of half to three quarters of an hour.'[8]

The equipment used for elver fishing in the Severn is simple, for apart from a fine-meshed, home-made hand net, the fisherman requires a bucket in which to empty the catch, a lamp, two forked stakes, known as 'tealing sticks', and a sack to carry the catch. Taylor describes the elver net as follows:

'The Gloucester elver net is a simple home-made implement of great simplicity but considerable efficiency. The handle, or "tailstick", varies from seven feet six inches to nine feet, according to the user's fancy. It passes through the headboard at an angle and reaches to the beginning of the curve in the top of the net. The headboard is approximately semi-circular in shape, sometimes of solid wood (elm), but also often partly of wood and partly of netting supported by a hoop. In the latter case the wooden section measures fourteen inches by five inches. The sidesticks of "red withy" from thirty-four inches to thirty-eight inches in length, are housed in notches or holes cut in the forward corners of the headboard and here they are halved, bent and secured with nails. Two strainers of wild rose briar pass through holes in the rear of the headboard, but are not here secured, curve together to pass behind the top of the tailstick where they are lashed, and then spread again to meet the junction of the sidesticks and headstick. The latter is about twenty-two inches in length and again of "red withy". A net of these dimensions requires a yard and a half of netting thirty-six inches wide. The traditional netting material is an undyed linen scrim, but in recent years both terylene and nylon have come into use for this purpose.

'Below Gloucester the headboard of the net is more usually triangular in shape with a rounded apex and made entirely of wood, but in other respects these nets are of similar construction to those in the Gloucester district. The reason for the difference is that below Gloucester the nets are more often used in shallow water, and the lower edge of the headboard is allowed to rest squarely on the river bed.'

The tealing sticks are used when the run of elvers is poor in order to peg down the handle of the net so as to place in the correct position for fishing. The net is therefore left in position for some time in order to catch elvers.

The Elver Fishing Act of 1876[9] specified that elvers could be taken on the Severn between 1 March and 25 April only, and thus centuries of controversey was ended. As far back as 1533 an Act[10] prohibited the taking of 'any young brood, spawn or fry of eels'. Although this was repealed by an Act of 1778,[11] the Salmon Fishing Act of 1873[12] once more prohibited the taking of elvers. From 1876, however, elvers have been the only fish fry which may be legally caught as food.

Today, elvers are not caught in any quantity, except on the Severn, and they are usually cooked by frying in bacon fat. In the Chepstow district, where elvers were caught in the Wye in the nineteenth century, the elvers were 'scooped out of the river' (between Brockweir and Llandogo), 'washed, boiled and pressed in a colander into greenish-white cheeses. You cut slices from these cheeses and fry them like bacon.'[13] Camden in 1722[14] describes elvering in Somerset where 'little eels scarce so big as a goose quill . . . looking very white, they make them up into little cakes, which they fry, and so to eat them'. Waters says[15] that 'Once in the pail the elvers froth like newly drawn beer and a few inches of elvers will quickly form an inch of foam . . . cooking turns them from grey to white and on the table they look like spaghetti. They have a delicate flavour all their own yet people eat so many of them during their short season that they see the last of them for the year without regrets. At the beginning of May elvers begin to turn black, owing to the formation of bone in the fish, and are thenceforth considered unfit to eat until they may be caught as eels.'

Although the Severn estuary is the all important area for elver fishing, a certain amount is practised in other areas. In Cornwall, for example, elvers are trapped at the tidal weir on the River Tamar by the Cornwall River Authority, and the elvers are sent to a Middlesex trader for stocking Scottish lochs. The elvers are attracted into a moss-fitted trough by a small flow of water created by pumping. The elvers ascend the trough and fall into a holding tank from which they are removed for transportation.

Between 1966 and 1971 the weights of elvers caught at Gunnislake Weir were: 1966 – 274lb, 1967 – 310lb, 1968 – 672lb, 1969 – no run of elvers – trapping abandoned, 1970 – 320lb and 1971 – 224lb.[16]

EEL SPEARING

Eel spearing or 'stanging' was a very common method of eel capture until recent times and although it still remains a legal method of fishing in most parts of England and Wales, it is far less widely practised today than in the past. Section 1 of the Salmon and Freshwater Fisheries Act of 1923 prohibits the spearing of freshwater fish, but the same Act specifically exempts the eel from the designation 'freshwater fish'. The only river authority that specifically prohibits eel spearing is the Severn, where since 1911 an order of the River Board deemed the eel to be a freshwater fish that could not be speared.

This method of eel fishing differs from all others in that the eels are caught when lying in the mud. It is a simple method of eel capture that can be practised in daytime throughout the year, and is really the only method of catching eels in winter. Eel spearing is especially effective on muddy foreshores or shallow dykes, and spearing can be carried out from a boat, from a bank or from mud-flats after the receding tide. The spear is thrust downwards into the mud either haphazard or where holes in the ground indicate the presence of an eel in the mud. Unlike salmon 'leisters', flounder spears and pike spears that have barbed points to pierce and secure the fish, eel spears are made of flat metal tines usually with rounded ends set close together. They are designed in such a way as to hold the eel in between the tines without damaging it. In the past this was important, for processed or tanned eel skins were widely used in country districts for such tasks as making connecting swingles for threshing flails and for making gaiters.[17]

There are many dialect names for eel spears, the following being the most common:

Eel Shears (Shar)	Severn Valley, Kent, Sussex, Lancashire
Eel stichers[18]	Wiltshire, Lancashire
Stang[19]	Lincolnshire
Stang gad	Lincolnshire
Eel gad[20]	North Lincolnshire
Auger	Humberside
Pilger[21]	East Yorkshire, East Anglia

Gleve: Glaive	Suffolk, Cambridgeshire
Eel pick	Norfolk
Gaw	Lincolnshire
Pritch (Prick)	Fens

The eel spear consists of a wooden shaft, the length of which varies according to the depth of water to be fished, attached by a socket to an arrangement of iron or steel prongs that number from three to nine. In most cases, the edges of the leaf-shaped prongs are serrated or waved and are designed to hold the eel in between the prongs, while the spear is removed from the water or mud. Smaller spears, for use in streams and ditches, are usually equipped with ash handles from ten to fifteen feet long and can be used by men standing on a bank. In the Severn estuary, however, large eel spears equipped with ash handles as long as thirty feet are used. These spears, specifically referred to as 'shears', were unwieldy implements that could only be used from the decks of trows.

Basically, there are two distinct types, the broad relatively short spears designed for 'hard bottoms' and those designed for use in deep dykes and heavy clay and known to fishermen as 'spears for no bottom'. In the latter group penetration is vital; the spears are longer and narrower and each prong is less flattened than in the first type. Their shape is specifically designed to penetrate thick mud, where eels might lie, and usually they are not fitted with one or two bands of metal riveted at right angles to the prongs, as in spears designed for hard bottoms. In 'hard bottom' spears, each prong may be flat and in a single spear nine prongs are not unusual for wide coverage is of greater importance than penetration. One type of spear used in muddy water and dykes in East Anglia consists of an expanding head of four or six wires, twisted to a hook shape at the tips and placed on either side of a central flange. There is a cross-binding of twisted wire which can be adjusted according to the density of the mud. For really deep waters, an eel spear may be attached to a heavy ash handle as long as 20 feet or more, but for canals, ditches and dykes, short-handled and light spears are used. Charles Green[22] has attempted to classify eel spears on a regional basis into four basic types – Western, Kentish, North-eastern and South-eastern groups. Although within each group there are many variations, a distinct regional pattern may be identified.

In the West of England, the Midlands and Wales the characteristic spear, designated the 'Western group', 'consists of a socket welded to a plate ranging from a semicircular to a triangular shape, to which are riveted several tines arranged fan-wise. The upper ends of the tines are arranged so that the riveting is spread evenly over the whole plate'.[23] In some cases a metal strap may be riveted across the tines to give the spear added strength. The Kentish group, used in Romney Marsh and the Isle of Thanet amongst other places, has fewer than those of the Western group from Lincolnshire, Humberside and the Fens are also parallel tines. The socket and tines are made in one piece and the tines are not riveted to the socket. The so-called 'North-east group from Lincolnshire, Humberside and the Fens are also characterised by spears having sockets and tines in one piece, but the proportions are quite different from those of the Kentish group'.[24] Again the tines, numbering no more than four, are arranged fan-wise and the tines are shorter and broader than those of Kent. The group designated 'South-eastern' by Green show considerable variation in design, and although many of this type come from East Anglia, they are not unknown in other areas, often as far north as the shores of Morecambe Bay. The spears are more complex than the other three; indeed, some examples preserved in English museums may be modern, factory-made articles that could be bought from any fishing tackle dealer. A fairly common type, seen in a number of museums, consists of a socket, continued as a flat blade forming the central tine. The upper parts of the outer tines, usually numbering four, are bent inwards and secured to the central tines by rivets. Another common example consists of tines being made in pairs, bent to an arc and riveted to a socket. In these spears a metal strap is usually riveted across the tines to give the spear added strength.

According to Taylor[25] 'the earliest illustration of a true eel

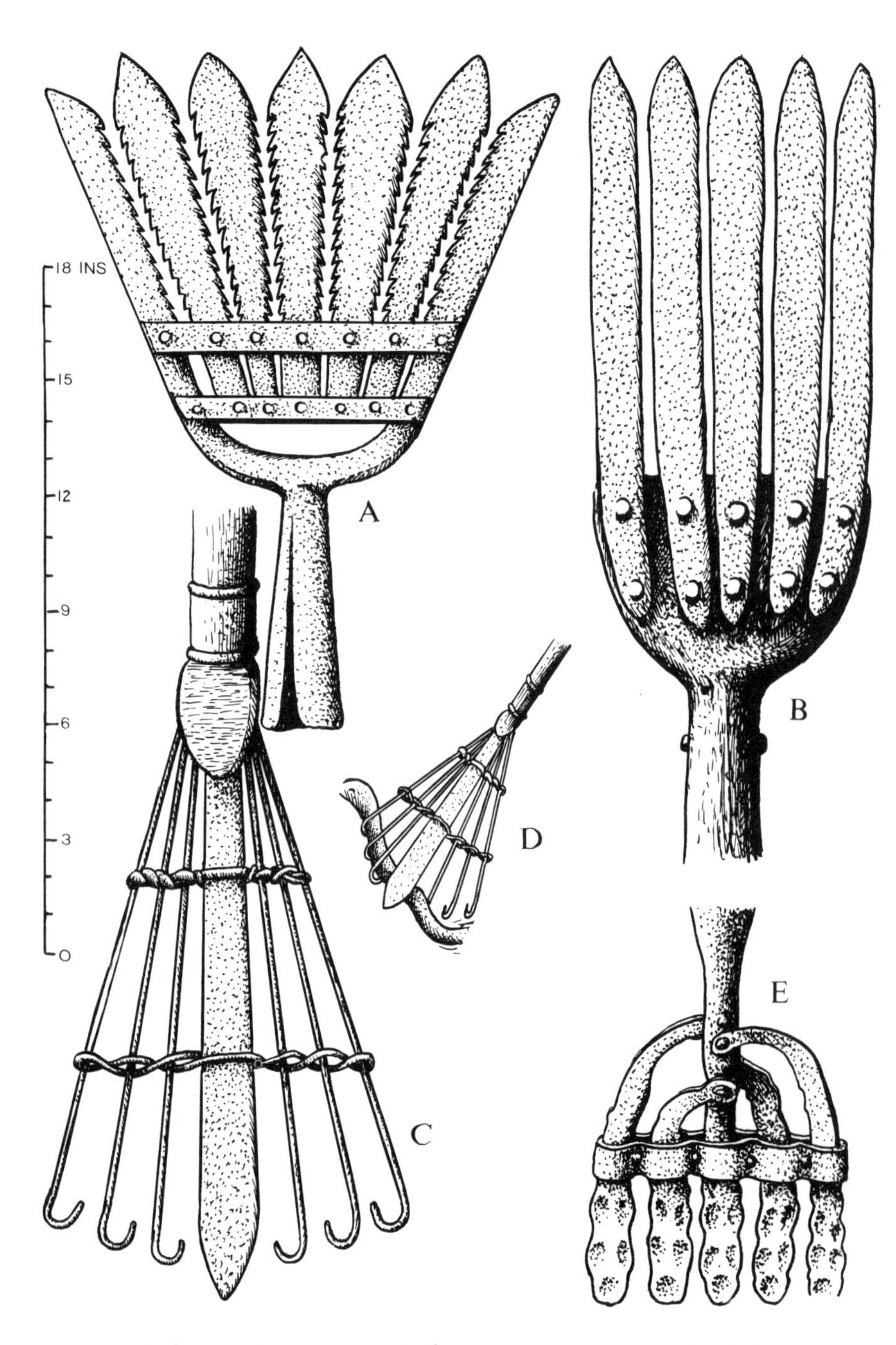

Eel spears (after Davis): A 'Hard bottom' spear (East Anglia); B 'No bottom' spear (Severn estuary); C Adjustable wire spear (Fens); D Use of adjustable wire spear; E 'Hard bottom' spear (Severn estuary)

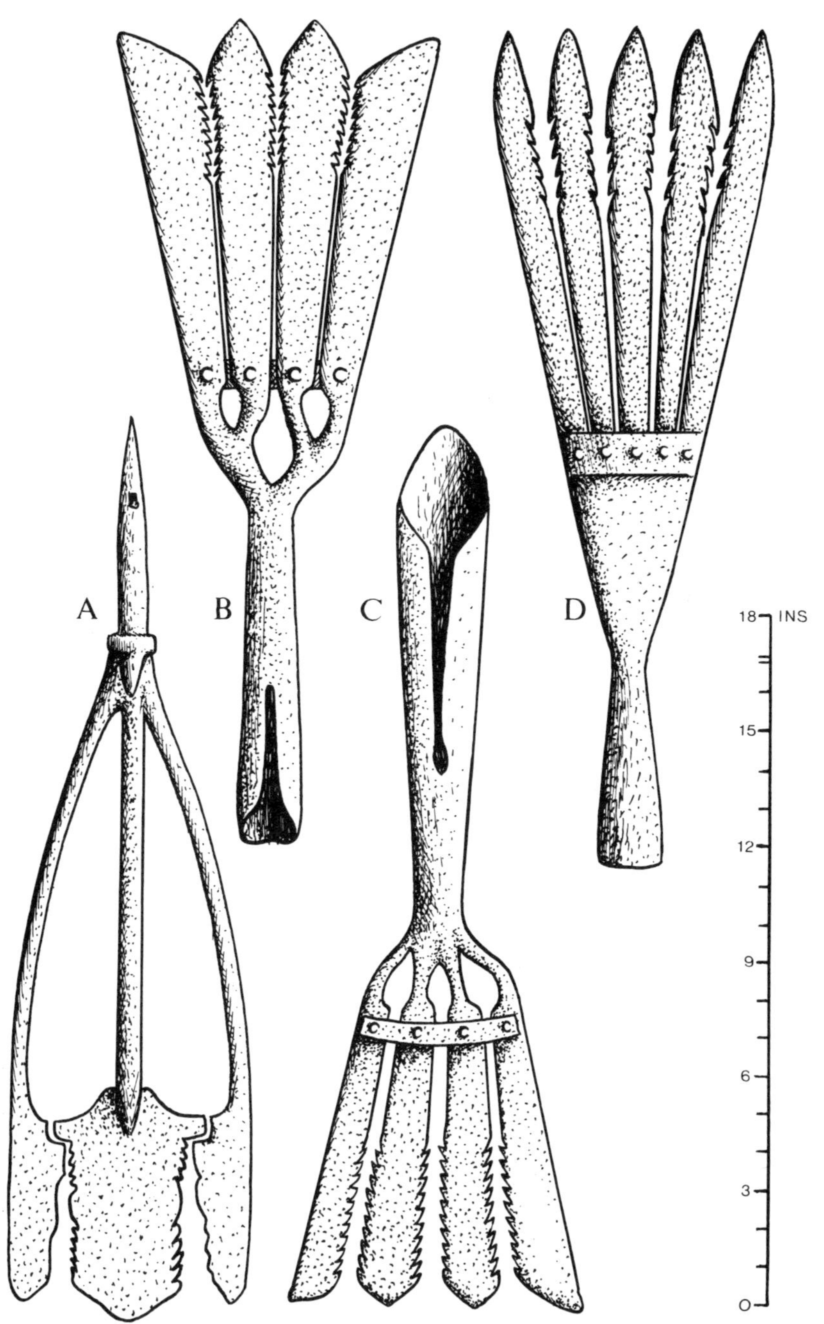

Eel spears: A Ribble valley, Lancashire; B Cambridgeshire; C Trent valley, Nottinghamshire; D Broads, Norfolk (made at Great Yarmouth)

spear appears in *Shirley's Book*, a late fifteenth-century MS of heraldic bearings. The Harding's coat, No 212, is blazoned argent, a chevron gules between three eel spears, points upwards, sable. One of the earliest actual descriptions of the mode of using an eel spear is given by John Gwillim in *A Display of Heraldry* (1st ed, 1611). Gwillim was a Minsterworth man and in his notes on the Arms of Stratele, he accurately describes the use of an eel spear.'

In the north-west of England, especially in the streams and rivers that run into Morecambe Bay, the method of eel spearing was somewhat different to the remainder of the country. J. McN. Dodgson, writing of his father's experiences in Preesall, Lancashire,[26] says: 'An eel spear was a 36 inch long iron shaft, ½ inch thick with a blade at one end and a ring at the other. Through the ring you threaded a clothes line with which to recover the spear after the throw. The blade was 6 inches long and divided, the inside edges being saw toothed, with the teeth pointing backwards. The eel-spear was thrown at the eel and was designed to bestride the body of the eel. When it attempted to wriggle clear, it worked itself on to the points of the saw-toothed jaw of the blade and became fast. The thrower stood by the water and watched for "a strike", a line of bubbles left by the creature as it swam. He aimed a little in front of the strike, and then hauled in on the clothes line.'

In winter, when pits and watercourses were frozen, many methods of eel catching were adopted in various parts of the country. In the Severn area, for example, eels were caught with short-handled tongs, the tongs being used for taking eels from the unfrozen margins of brooks. The tongs, of wood with metal teeth, usually measure between 12 inches and 15 inches in length. Similar tongs are also used for handling eels on Severn-side, while for the capture of eels in ditches, tongs attached to handles up to 8 feet in length were widely used. The tongs were closed by pulling a cord which closed the metal jaws of the implement.

The 'stitcher' or 'eel crook' is another instrument that was widely used in the ditches and streams of the Vale of Berkeley in particular. 'For stitchering', says J. N. Taylor, 'an old sickle roughly notched on its cutting edge and fitted to a long handle may be used in shallow ditches or ponds. A refinement consists of a specially made iron crook notched on its inner edges. When the nose of an eel is seen peeping from the mud bottom, the stitcher is used to flick the eel quickly on to the bank, where it may be caught by the hands.'

A method of eel fishing, known variously as 'patting', 'sniggling', 'tatting', 'bubbing' or 'clotting', is a universal and cheap method of eel capture. 'By means of a blunt darning needle, worms are threaded end to end, on a piece of worsted or twine, 2 or 3 yards long, which is then wound round the fingers and the loops bound together and fastened to the end of the line, which has a small sinker a short distance above the

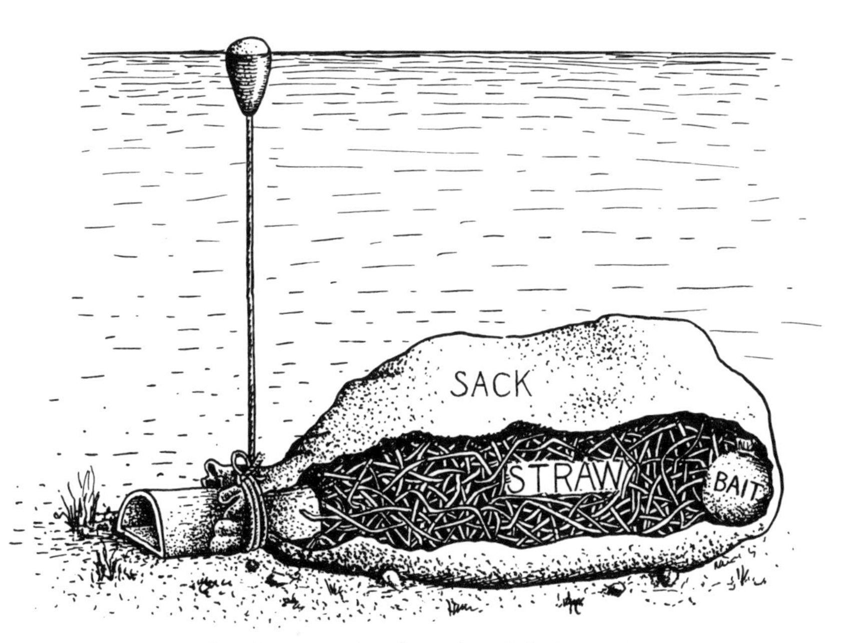

Simple eel trap made of a sack and drainage pipe

bunch of worms. The fisherman keeps the bait on or near the bottom and, on feeling a check or pull, waits a few seconds and then gradually and quietly lifts the eel out of the water; the eels hold on from greed or because their teeth are caught in the worsted.'[27] In West Wales this equipment, which was widely used in the past, is known as *llindag* and was regarded as a very efficient method of night fishing in the summer months.

In the Severn estuary, 'sniggling', as this method of fishing is called locally, is a widely used method of eel capture. 'Sniggling employs the ancient principle of the gorge', says Taylor,[28] 'and was described by Izaak Walton in *The Compleat Angler* in 1653. A length of twine is whipped to the centre of a stout needle and the latter threaded longitudinally through a fresh worm. The point of the needle is stuck lightly into a stick and the sniggle pushed as far as possible into a hold or bank undercut, for such are the favourite haunts of eels in daytime. Presently the eel will take the worm. The fisher, allowing time for the eel to swallow the worm and needle, then begins to pull on the twine. The needle crosses the gullet of the eel, which eventually tires and may be drawn from its hole.'

Sniggling in other parts of the country, especially in north-west England, is done by another method.[29] In icy weather, a hole is made in the ice on a pool and a bundle of straw at the end of a rope is dropped through. This is left in place for an hour or two or overnight. The bundle of straw is then hauled out on to the bank and the eels that would have gone into the straw while it was submerged are landed.

Straw is also used for another simple method of eel trapping, more especially in the Fens and East Anglia. A bait in the form of fresh-water mussels with shells crushed, lamperns, gudgeon, small fish or offal is placed in the bottom of a sack and the sack filled with straw. A drain pipe is attached to the mouth of the sack and a marker tied with string to the trap. The whole contraption is then sunk to the bed of a stream or pond. In baiting traps, it is essential that the bait is enclosed in woven or perforated material, so that the scent may be diffused through the water. The whole trap is removed at regular

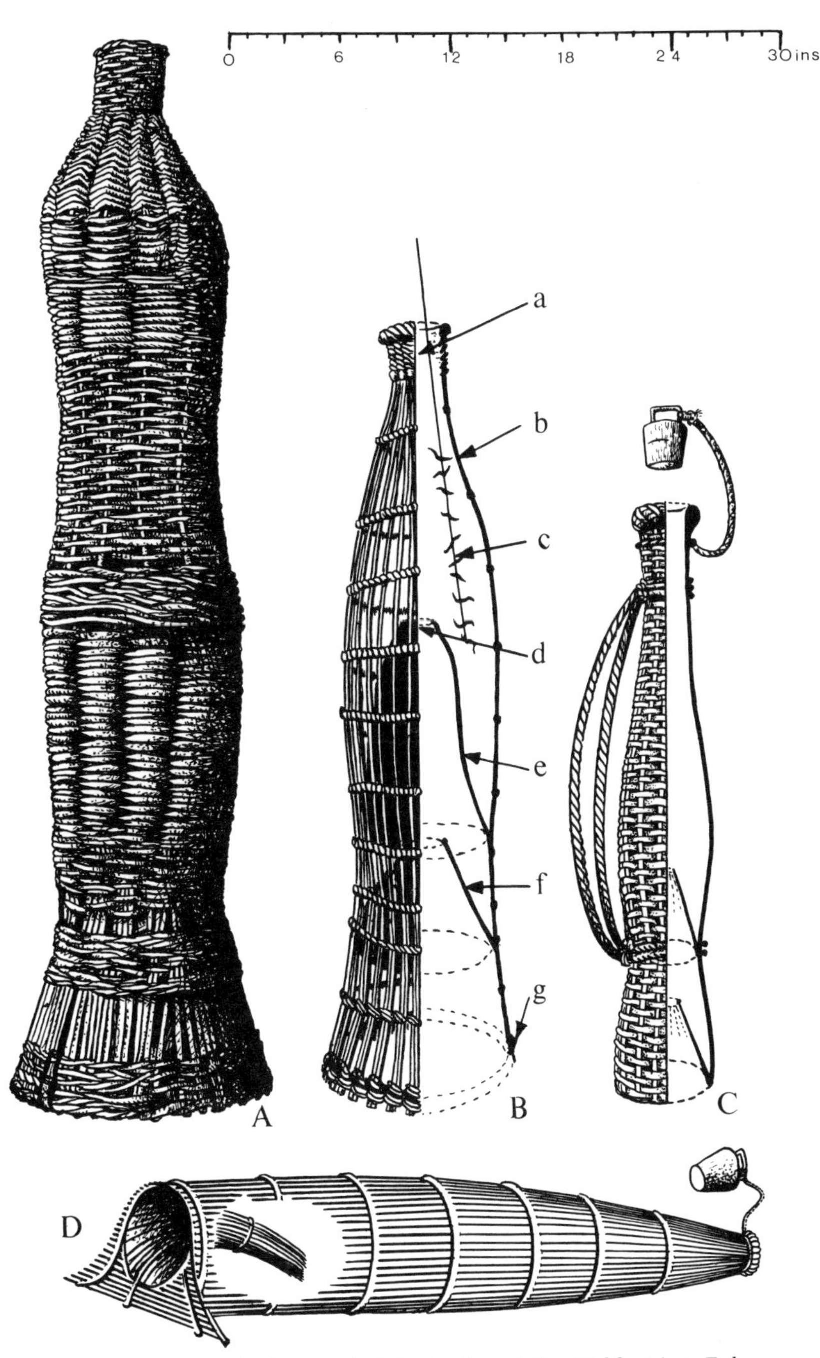

Varieties of eel pots: A Eel grig, Fens (after Tebbutt); B Eel trap, Severn, Worcestershire (after Peate) (a) Starling closed with grass (b) Cod (c) Bait (d) Loop (e) Middle inchin (f) Far inchin (g) Head; C Eel hive, Fens (after Porter); D Eel basket, East Anglia (after Davis)

intervals and the eels caught up in the straw removed.

PORTABLE TRAPS

Portable eel traps or 'eel pots' of wicker-work or metal were widely used for the capture of eels in all parts of England and Wales, but more especially in the Severn Valley, Lincolnshire and the Fens. They are known variously as 'grigs' (East Anglia and the Fens), 'kiddles' (Thames estuary), 'hives' (Lincolnshire and Cambridgeshire), 'putcheons' and 'wheels' (lower Severn Valley) and 'wills' (upper Severn Valley). Eel pots are long and narrow with one or two non-return valves inside. They are set, usually singly, in a stream through the opening that falls up-stream. The eels make their way to the front of the pot and fail to find a way out. By removing the wooden or cloth bung at the front of the trap, the eels may be removed. In shape, eel baskets have been described as 'roughly resembling a closed umbrella'[30] or 'shaped like the straw covering of a wine bottle, being some three feet long and baited at its end with a piece of lamprey or rabbit'.[31]

In the Fens, eel pots, known locally as 'grigs', are usually made of closely woven withies and they are of slimmer appearance than those of other districts.[32] Each grig measures up to 60 inches long and is bottle shaped. The entrance, facing up-stream when the grig is in use, is known as the 'head' and the narrow tip as the 'bottom end'. The two cones of pointed sticks are known as 'chairs'. In addition to this fairly common type of grig used in the Fens, there is another type of smaller trap known as a 'hive'. The eels enter the first chair directly, the cone being woven straight from the entry to the trap. Beyond the second chair in the widest part of the trap a bait, usually a worm, is attached to a piece of wire. In the Fens, the large-sized grigs are not always baited, for in setting a trap it is inserted in a hole in the middle of a net stretched across the narrow part of a water-course.

The hive, on the other hand, was always in the path of the eels' annual autumn migration to the sea. When the current is too strong for wicker traps, wire ones take their place, but although metal traps are more durable, neater and easier to construct than wicker pots, many eel fishermen believe that eels dislike metal and that traditional basket traps are far better for their capture. Nevertheless, metal and wire traps are frequently found in various English rivers, particularly those of eastern England. Double-entry wire traps, similar in many respects to a lobster pot, are widely used. They are, however, more elongated than lobster pots and the inner opening of the funnel much smaller, so small, in fact, that the eel can only just squeeze through it. While in the Wash, the eel grigs or 'butts' are net cylinders supported on hoops, in south Lincolnshire wire pots are preferred. The pot consists of a wire mesh stretched over a cylindrical metal frame, consisting of six or more hoops, with a pair of metal bars on either side connecting the hoops. The trap has two non-return valves, of wire, like the wickerwork pots, but is equipped with a wooden end door at the front. This is hinged and kept in place with a short metal bar at the bottom. In Norfolk, especially on the Broads, eel catching is widely practised and baskets were widely used until recently in the ditches and drainage canals that abound in the district. A Broads trap which shows a distinct bulge in its centre is made up of a large number of longitudinal green withies placed very close together and bound with half a dozen willow loops. Both entry and top are strengthened by a woven loop of withies. Most Broads traps are equipped with only one non-return valve which is very simply constructed. This consists of pieces of withy rod fixed to the entry and converging inside at the widest part of the trap into almost a cone-shape inside the trap. These yield to pressure and allow the eels to enter the trap.

On the upper Severn, on the Montgomeryshire-Shropshire borders, 'grigs' or 'wills' were common about Shrawardine. ' . . . The fishermen stretched a net across a ditch as a flood was subsiding. The mouth of the will was placed in a hole in the net. The only way, therefore, for the eels to avoid the net in order to continue their intended way down stream was for them to enter the will.'[33] Eels entering the pot do so through a narrow opening made of inwardly directed osier spikes. These yield to

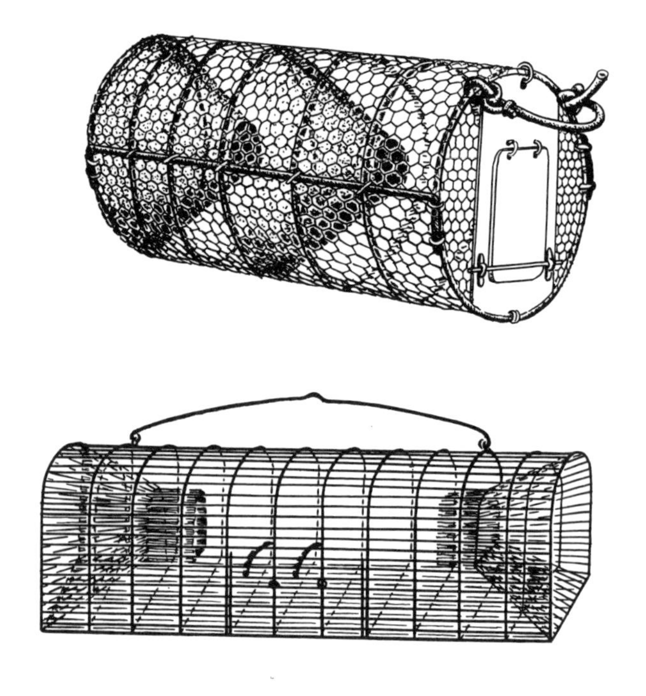

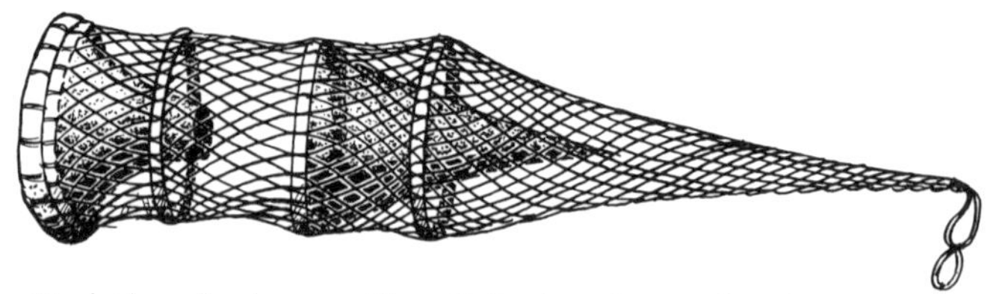

Varieties of eel traps: Top Eel grig, wire mesh, Lincolnshire; Middle Double-entry wire trap; Bottom Eel net (Dutch Fyke net)

pressure and let the eels enter, but come together again and prevent their return.

On the Severn, the eel baskets are not as closely woven as those of eastern England. In Fenland grigs, for example, the withies are closely interlaced in the manner of a true basket, but on the Severn the round withies arranged lengthways are bound together by withies that encircle the pot at 4 inch or 5 inch intervals. The wicker-work has the appearance therefore of being far more open. Describing the eel pots of the Worcestershire Severn, Peate says[34] that 'The Severn eel-trap is a basket 44 inches long . . . made of withies bound skilfully together. Inside the trap are fixed two cones of split splints or "stubbs" made of about twenty-four split rods each. The stubbs have sharp pointed ends and the two cones are known as "the far inchin" and "middle inchin". The large opening into the trap is called the "head" and the other end the "starling". The middle inchin, which is the inchin furthest from the head, has a much smaller aperture than that of the far inchin. Over this aperture is placed a loop which prevents the stubbs from being opened either by the trapped eels or through the action of water. . . Two or three small osiers are peeled, a few leaves being left on. These are then covered with garden worms fixed on the osiers and are placed through the starling, in the "cod" . . . The starling is closed with a pad of grass or hay. The trap is placed with the head downstream, the starling being fixed with a line attached to a stake or to a bush on the bank and the head held in position by small stones or bricks tied to the basket. The starling is also weighted in the same way. The eels worming their way through the inchin to the cod find that their return is prevented by the sharp ends of the stubbs, which open for their ingress, but prevent their egress, the loops preventing the splints from being forced open even if the sharp points be overcome.'

In the lower Severn, below Tewkesbury, two sizes of basketwork eel traps are used: the putcheon and the larger wheel. A putcheon usually measures about 40 inches long with a head about 10 inches in diameter. The wheel, on the other hand, may be a 'half wheel' at least 50 inches in length and about 14 inches in diameter at the mouth or it may be 'a wheel' proper, which is larger again. Both putcheon and wheel are 'shaped in the same way with a wide opening at one end, bellied in the centre and tapering to a small opening at the other end. Inside the trap are fitted one, or most usually two, constricted throats of canes directed towards the narrow end'.[35] These constrictions are called 'chales' and they are made separately on wooden chale blocks to a near conical shape. These are then braided into position during the making of the main body of the trap. A wooden plug or piece of rag is used to block up the narrow end of the trap which will contain a piece of rabbit, shrimp or lamprey as a bait. Around Oldbury on Severn, however, the wheel is regarded as the smaller type of trap, each one being approximately 24 inches long while the larger basket, up to 54 inches long, is termed a putcheon. Shrimp baits are preferred in this area.

It is interesting to note that lampreys, that were in the past widely eaten by Severnside dwellers, are used today for eel bait and are hardly ever eaten. Although lampreys may be caught while netting for salmon, it is said that the fish are bruised by this method and those caught in basket traps, similar to eel putcheons, are preferred. 'During the last century', says Taylor,[36] 'considerable quantities were sent to east coast ports for use as cod bait. They were caught by means of basket traps called "lampern wheels", staked in bundles at the river's edge.' These wheels were similar in size and method of construction to the more common eel wheels with chales inside. For storing lampreys, large urn-shaped baskets, known as 'hard wheels' or 'cunning kipes' and stopped with a wooden plug, were used to keep the fish alive. Two examples at Bishop Hooper's Lodging Folk Museum, Gloucester, measure 19 inches and 38 inches in height and 52 inches and 60 inches in diameter respectively. 'A tethering rope is thrown around the neck of the basket and the whole sunk in the river, where the fish remain alive until required. Similar baskets were used for the transport of lamperns to the east coast, and a man had to travel with them to stir the contents or the fish became smothered.'[37]

Undoubtedly eel baskets are of considerable antiquity and examples are illustrated in the fourteenth-century *Luttrell Psalter*[38] and the sixteenth-century Peniarth manuscripts,[39] for instance. In the *Luttrell Psalter* the conical-shaped traps are shown in position in a mill race; indeed many mills incorporated eel traps in streams until the late nineteenth-century. Many complaints were made to the Commissioners on Salmon Fisheries in the early eighteen-sixties that traps designed for eels in mill races were often catching young salmon as well. 'We believe', says their Report,[40] 'that these contrivances have been set up without any warrant or title . . . they form a standing violation of the principle, if not the letter of the law; and we are convinced that if allowed to carry on their destructive operations as heretofore, all other measures that may be taken for the increase of the supply of salmon will be unavailing.'

In addition to portable traps, fixed traps are still used in some parts of the country as, for example, near Llangorse Lake in Brecknockshire. These traps are usually at mills or at the outlets of lakes. The essential feature of the automatic trap is that the water falls through a grating which intercepts the eels, the grating being set between two walls built of timber, brick or masonry.

EEL NETTING

Of all methods of eel capture practised in England and Wales at the present time, by far the most common are conical eel nets. These nets consist of 'a long, small-meshed conical bag, inside of which are one or more non-return valves, with the wide end towards the mouth of the net and tapering down to quite a small opening behind'.[41]

The stronghold of commercial eel netting is the eastern counties of England, between the Thames and the Humber. No figures for the total catches are issued, but they are substantial and a large number of licences for eel fishing are issued by the River Authorities in the eastern counties. In the area administered by the East Suffolk and Norfolk River Authority, for example, 590 eel fishing licences were issued in 1971, and of this total, no fewer than 457 were for fishing with the so-called Dutch fyke net.[42] Further north, the Lincolnshire River Authority issued 439 fyke net licences, while the Welland and Nene River Authority issued 166 licences and the Great Ouse River Authority 14. Outside the eastern England area, eel netting is not so important, although the Severn River Authority issued 43 licences for 'nets and putcheons'.

Funnel nets for eel catching may be divided into three types, 'the coghills', 'Dutch fyke nets' and 'wing nets'. The simplest form of net is the coghill, widely used at present in parts of Ireland. These are 'tapering nets about 20 feet long with the mesh very small at the rail end and increasing in size towards the mouth; at some distance from the end is a short conical valve net, open at both ends, which prevents eels that pass it from returning. This inner net is kept in shape by a pair of hoops fastened inside the main net, one at its attachment, the other nearer the rail end. At its mouth, which is about 8 feet square, the coghill is fastened to two upright poles, each of which has an iron ring secured to it near the lower end; when the net is to be set, these rings slide down the bars, which are secured at their lower ends to the inner face of the uprights; the upper ends of the poles are then made fast.'[43] The coghill has no wings or leaders to guide eels into the net.

As far as can be ascertained, the true coghill was never in use in England and Wales but closely related to it, though rather more complex in construction, are the wing and fyke nets that are widely used in eastern England. The catching part of both wing and fyke nets is the same. 'It consists of a conical net, usually made of knotted or knotless nylon nowadays, with a series of hoops diminishing in size from the entrance to the cod-end, fitted inside it to keep it open. Two or three funnels or valves of the same mesh netting are fitted inside so that eels entering the net do not swim out again. Wing nets differ from fyke nets in that they have two walls of netting attached to the hoop at the entrance to the net to divert eels into the trap whereas fyke nets have only one. Wing nets are usually set facing up or downstream in slow-flowing rivers and creeks; fyke nets athwart the stream. The two walls of netting are attached to the sides of the entrance hoop in wing nets; the single wall known as a ledder or skirt is attached vertically across the centre of the entrance hoop in fyke nets so that eels passing either up or downstream are diverted into the net.'[44]

In using a wing net, the wings and net are attached to stakes firmly driven into the bed of the stream. 'It is important that the lower ends of the wing should be close against the ground; a leaded foot-rope helps to achieve this, and a second leaded foot-rope about six inches from the edge of the net gives additional security against the eels burrowing under it.'[45] The non-return valves inside the nets are to prevent the eels from swimming backwards to the mouth of the net. The mesh of a wing net is much finer at the apex than at the entrance to the net, for, says one writer,[46] 'it is unwise to have all the meshes of the smallest . . . It would appear that a sudden checking of a rush of water at the mouth of a narrow *cul-de-sac* tends to produce backwash and eddy currents, thus preventing the entry of the maximum of water, and facilitating the escape of some of the catch, and also the closer and smaller the meshes of the net, the more resistance is offered to the stream and . . . the whole apparatus may be carried away.' Wing nets may be set singly across the flow of a narrow stream, or they may be set in combinations, their arrangement and number varying according to local conditions. On the coast and in wide estuaries, for example, each net faces in the opposite direction to the next. In some cases woven wattle hurdles may replace net wings, while in some rivers in the spring and early summer the nets may be set with their mouths downstream to intercept eels as they make their way upstream. Most eel fishermen do not use bait of any sort in an eel net; indeed, baiting is said to be undesirable for it tends to divert the eels away from the wings to the outside of the nets.

Although most wing nets measure no more than 20 feet in length, in the Severn estuary nets as long as 50 feet are used, especially on the Llanthony Severn, just below the city of Gloucester. 'A full sized net because of its finer mesh, takes three months to knit and is immensely heavy, yet an eel net will last a fisherman twenty seasons, for the migration of eels, passing down river at the rate of nine and a half miles a day, allows for little more than seven or eight nights in the year in which eels may be caught by this method.'[47] In the Severn estuary, the force of the current is such that wing nets have to be anchored with chains and strong ropes.

The fyke net is also a conical filter net, but in this case a single width of net attached to the first hoop of the cone and in its centre acts as a leader to guide the eels into the net. The so-called 'Plough Net' of Boston, Lincolnshire, is of this type, and usually measures 19 feet in length. 'The last 13 feet are known as "the gear"; the meshes at first are of such a size as will not necessarily retain the eels, and the net is gradually graded down to a mesh that will retain the fish. The gear is supported by hoops, which also support a series of funnels, 2 feet long (locally "inchers" or "chairs").'[48] When Davis was writing in 1923, plough nets were far less common than wing nets, but since 1945 they have become the usual method of eel capture in eastern England. Since that date too, they have become known as 'Dutch fyke nets', undoubtedly due to the presence of Dutch eel fishermen in the rivers of East Anglia and Lincolnshire. Until 1968, for example, approximately fifty eel fishermen from Holland used to pay extended visits to the tidal River Hull using fyke nets for the capture of eels.

The fyke nets used by eel fishermen today are produced commercially by a Bridport, Dorset, net manufacturer to a standard size. The net measures 12 feet long with a 15 foot leader attached to the entrance hoop. The leader is 2 feet high and has four plastic floats spaced 3 feet apart along its head-line and half a dozen cylindrical leads on its foot-rope. The conical net itself has seven compartments separated by cane hoops of varying diameter, and a cod end. The entrance hoop has a 21 inch diameter, and the second fitted 21 inches from it is 17 inches in diameter. 'Thereafter hoops of 16, 14, 12, 10, 9 and 8 inches diameter are fitted at distances of 17, 16, 14, 12 and 14 inches apart, followed by a cod end 24 inches long. Funnels are fitted between hoops 1 and 2, 3 and 4, 5 and 6.'[49] Fyke nets are staked with two or more stout stakes, and they are, where currents are strong, also anchored in place across the flow of water.

In some districts, especially on the northern side of the Thames estuary in Essex, trawling for eels is widely practised, especially where there is a glut of eels during summer nights. Beam trawls, 30 to 50 feet long, are used for this. The beam trawl is a conical net with the upper side of its mouth fitted to a wood or iron beam and iron heads or shoes which keep the beam off the river or sea bed. A cylinder of small-meshed net is attached to the main core of the net a few feet from its apex and forms a valve and cod end. The beam for an eel net may be from 14 feet to 20 feet long and the heads are usually of wood. In use the cod end, which may be as long as 20 feet, balloons out around the main net.

For notes to Chapter 10, see pages 323-4

11 Poaching

Poaching may be defined as taking, or attempting to take, fish by means of an unlicensed instrument other than rod and line, fished with a conventional bait. It includes fishing during the closed season and the use of nets and other instruments, by persons not licensed to use those instruments. In Welsh rivers during the year March 1970-March 1971 no fewer than 518 offenders were prosecuted for poaching offences, but actual prosecutions must have represented only a small proportion of actual offences.[1]

Until the early part of the present century, the taking of ripe salmon off the spawning beds at a time when they are particularly vulnerable was regarded as a necessity of life in rural communities in order to provide food for the winter months. Salmon was salted and smoked and it was commonplace in the humdrum diet of riverside communities. The roe of the female salmon was prepared to make 'salmon paste' or 'jam' and illegal bait for trout, which was sold to unscrupulous anglers and helped to eke out the low wages of the country dwellers. The improved economic status of rural communities during the present century has obviated the need to rely on poached salmon to supplement the winter rations, but the additional 'pocket money' which the sale of roe provides is still an incentive to go poaching. 'A 10lb salmon caught in May', says Millichamp, 'could fetch only 50 pence a pound in a glut year; that is £5. The same fish if it survived until the spawning season would weigh approximately 9½lb and would produce around 2½lb of roe. This as jam, would fetch up to £3 per pound, that is £7.50; and the carcase could fetch 25 pence per pound, that is £1.75. Therefore the fish would fetch in total £9.25 or almost twice the price it would fetch in May!'

Some indication of the prevalence of poaching in the mid-nineteenth century is provided by the *Minutes of Evidence taken before the Commissioners appointed to inquire into Salmon Fisheries* in 1861. On some rivers poaching had reached mammoth proportions and the stock of salmon was becoming seriously depleted as the result of poaching activities. On the Tywi, for example, 'the closed time' was 'not observed at all . . . the men fished night and day, summer and winter' with the result that the river was becoming a poor salmon river.[2] Illegal nets, spears, gaffs and poison were all used by hundreds of offenders. In North Wales the Dee was regarded as being particularly haunted by poachers, for, said one witness,[3] 'it grieved my heart for a long time to see the abominable waste and destruction of salmon in the Dee. Spawn is taken away by the farmers for their pigs, poultry and ducks and even as manure to their land . . . men with corracles on their backs . . . taking all they could catch with very fine nets. They all met at a certain public house, where large wicker baskets were filled with what they had obtained and sent to the Liverpool and Manchester markets . . . They fixed stake nets . . . and there were miles of those stake nets. We appointed Mr Hill, the head of the Chester police, as our conservator and as it was absolutely needful that these stake nets should be done away with, he was ordered to take such a force as was requisite and have them cut down and removed . . . Below Erbistock Weir people came with four or five hooks attached to strings and pulled them out.' The position was similar on all other rivers and as the Introduction to the Report says:[4] 'The modes of destruction are various . . . Killing of breeding fish at the spawning grounds is a very prevalent abuse. On the upper parts of the Wye it is carried on with a high hand by large parties of men setting the law at defiance, and acting in such a manner as to strike terror and overawe resistance.

'On numerous rivers the practice of going out at night with spears and torches, and what is called "burning the water", is in full force. Elsewhere the killing of spawning fish goes on less openly, but with very pernicious effect. They are taken with nets, gaffs and instruments of various kinds, at the foot of weirs and other obstructions where they are detained in their passage, and offer an easy prey to poachers and marauders.

'The killing of spent fish in the early part of the year is another abuse by which an injury is committed against the public, to which the advantage gained by those who do the mischief bears no proportion. To destroy a salmon in this stage when it is of scarcely any value for food or sale, while, if it be allowed to return to the sea, it would in a short time become a valuable fish, is another instance of that short-sighted cupidity which it is in the interest of the community, and should be the office of the law, to check. At present the law is quite inefficient for the purpose, and in some instances the too early termination of the close time (several of the rivers opening in January, when clean fish are rarely to be met with), offers a temptation to kill the fish wherever any means of disposing of them are to be found.

'Another great source of the depression, brought prominently before us at various places, and regarded by many as a main cause of the decrease of the breed, is the wholesale destruction of the salmon fry. In illustrations of this statement, we would refer to the evidence given to us on the Wye, the Teifi, Tees, Ribble, Lune, Test, etc.

'The young salmon at this stage of their growth variously denominated, according to the provincial terms, "last-springs", "samlets", "skirlings", "brandlings", etc., etc., are killed in great numbers; in some places by the rod and line, in others by nets, baskets, and various contrivances used at the mills, where they are trapped in their passage down. Being esteemed as a delicacy for the table, they are hawked about and sold with little disguise in the towns, and supplied by the innkeepers to their guests, though the law distinctly forbids the capture of them. We found many persons disposed to make light of this abuse, on the ground that the extraordinary fecundity of the salmon tribe renders such a loss of the fry, though wrong in principle, insignificant in effect. We are satisfied however, that these persons form an erroneous estimate of the waste of life thus caused, and that the protection of the fry, which it has been the object of the salmon laws from early times to enforce, is essentially necessary to restore the rivers to their natural productiveness. When it is considered within how short a time this small and comparatively valueless stock would, if allowed to pass unmolested to the sea, return in greatly increased size and good condition, the destruction of the fry must be regarded as another instance to be added to those already mentioned, of a wanton waste of valuable food.

'With reference to the capture of unseasonable fish, whether in a spent or spawning state, when they are unsuitable, if not unwholesome, for food, the question naturally occurs, What causes the demand for them, and how are they disposed of? Assuredly, if there were none to purchase such fish, there would be none to take them; consequently, a market must exist somewhere. It is difficult to trace the channels of a contraband trade, which the parties who carry it on are interested in concealing, but we were able to obtain some facts affording a clue to the outlets of this traffic, which, if it could be followed out, would lead, we believe, to singular disclosures. We have reason to suppose that much of the unseasonable fish caught in different parts of the kingdom, is sent to London, consigned to large dealers in this commodity. A certain quantity of it is afterwards dried or "kippered", a process which in a great degree disguises the original quality of the article.

'But in addition to this outlet we received from various quarters information to which we attach credit, that a great proportion of the foul and unwholesome salmon taken out of season finds a ready market in France. We were told by a person in Northumberland, who himself deals largely in fish, that there are agents for the fish dealers of Paris on the north-east coast of England, through whom the traffic is carried on without the least difficulty and to a considerable extent. The price paid for these ill-conditioned fish is small; it was stated about 6d a pound. The ingenuity of French cookery succeeds, we may presume, in making palatable that which in its natural state would be both distasteful and injurious. The export to France has, we understand, much increased since the import duty on salmon was taken off in the year 1856 by the French Government. We have obtained a return from the English Customs' Department, showing the declared value of all the salmon exported from the principal ports under that designation during the last three years. In 1858 the value exported to France was 4,539 l.; in 1859, 3,469 l.; in 1860, 10,746 l. But this return shows the legitimate trade only. We have no means of ascertaining the quantity of unseasonable fish sent abroad by various routes and conveyances, and which purposely avoids observation. We believe it, however, to be very considerable. The evidence given us by persons connected with Billingsgate Market shows that it amounts to a great quantity. The clerk to that market stated that on some days the quantity sent had amounted to as much as three tons. This traffic undoubtedly operates as an encouragement to illegal fishing, and ought therefore, to be checked by all practicable means.'

As a result of legislation in the eighteen-sixties and seventies, more and more Boards of Conservators were set up for the various rivers and these made an all-out attempt to stamp out poaching. Nevertheless all attempts have failed and poaching still presents a major problem to the river authorities. 'On a small salmon river', says one writer in 1968,[5] 'perhaps a thousand fish would make it to the headwaters. Of them, according to one Fisheries Officer, possibly 80 per cent could be wiped out by casual poachers working as individuals . . . oddly enough the spawning salmon seem scarcely worth having. They are nothing like the handsome creatures that entered the river in spring. The flesh is white and flaccid, not to be bought by the most credulous and inexperienced hotel keeper. A long time ago in hungrier days these salmon were undoubtedly taken to feed families, unappetizing as they undoubtedly were.' It is considered important that salmon is caught on its way to its redd and before it lays its eggs, for after laying the fish become tough and almost inedible.

The high price of salmon in recent years has given rise to a new type of poacher – the professional or semi-professional. 'These people', says Millichamp, 'care little for conservation and often use methods which result in the mass destruction, not only of salmon and trout but every other living thing over several miles of river. They are usually organised, sometimes operating in gangs, and cover large areas of country. This type of poacher does not confine his activities to any one period of the year; at no time are fish safe from his predations.'

In addition to poaching on a large scale for profit, another factor that has contributed to the widespread occurrence of illegal fishing in recent years has been the acquisition of fishing rights on many rivers by outside persons. Stretches of river, regarded from time immemorial as 'the property' of local anglers, have been bought up by individuals and syndicates at fantastically high prices. This has led in some instances to local people, who have lost their rights to fish on their 'own' river, to deliberately set out to take as many fish as possible, by any means – fair or foul – from the fishing concerned.

Traditionally a great variety of ingenious instruments have been used for the illegal capture of salmon in the rivers of England and Wales. The Fishery Report of the eighteen-sixties, for example, noted that one of the most common methods of salmon capture was the 'bushing' of fish. By this method a bush, a hole or rock was surrounded with a fine-meshed net and the salmon driven into the net by beating.[6] Another writer, about 1910, noted that 'bushing' was widely practised on the Usk and the other rivers of Brecknockshire:[7] 'This very killing practice is much followed here . . . The "old hands" know exactly where the fish lie. Two or more of them go together, well laden with stones, and commence pelting the "catch" most vigorously, until Master Salmo finds it convenient and prudent to retire, which he does by swimming to the sides and taking shelter under a root or bush; when the commotion subsides a little, the "old hand" goes to the most likely bush, lies down with his face as near to the water as possible, and earnestly scrutinizes every nook and corner, frequently taking off his hat and moving it backwards and forwards to aid his vision by throwing a shadow on the water; presently he espies it, and if his companion is not prepared with a good hazel-stick it is very quickly procured;

a "gaff", which is always carried in the pocket, is firmly tied on with strong cord, which cord is always attached to the gaff; a survey is now made to see that the coast is clear, and in another minute the fish is writhing on terra firma. Two or three blows on the head, and a strong cord put through his gills with a large loop in it is the work of a moment; then one of the party "doffs" his coat, puts the loop over his head, the salmon of twelve or fifteen pounds hanging down his back, which is quickly covered again by putting the coat on. If either of them is licensed, he carries a fishing-rod in one hand, the salmon in the other, and marches triumphantly home; and who is to say he was not taken by the fly in fair angling? If not licensed they always separate, and go home in different directions; the fish is then hawked about the town, and sold at once for a shilling or two the pound. The proceeds are then equally divided, the old rendezvous is sought that night, and over many a quart of "cwrw" they enjoy their triumph over the keepers, and when the money is all gone, which is mostly the case the first night, they drink, again and again, to their next "merry bushing".'

The single-barbed gaff seems to be a universal implement for the capture of salmon. Each one is in the form of a metal hook up to 12 inches in length with a sharp barb on the inner side of the tip. Some of these gaffs are permanently fixed to a wooden handle, either with string or a socket; some may be disguised as walking or thumb sticks, but most have removable hooks that can be snapped onto the wooden handles with considerable speed, when the need arises. Although gaffs may be used in daylight, they are more often used at night in conjunction with artificial light.

'Burning the water' seems to be a widespread and universal practice and Alexander Fenton in a recent paper on Scottish Salmon Fishing Spears, for example,[8] describes how bunches of heather dried in the kiln of a corn mill, tarred sacking, birch bark, the staves of old barrels, sods soaked in paraffin and dried broom or fir tops were all used to provide a blaze. In Wales, however, a light was provided by holding above the water a sack, soaked in pitch, or oil on a stick. One informant referring

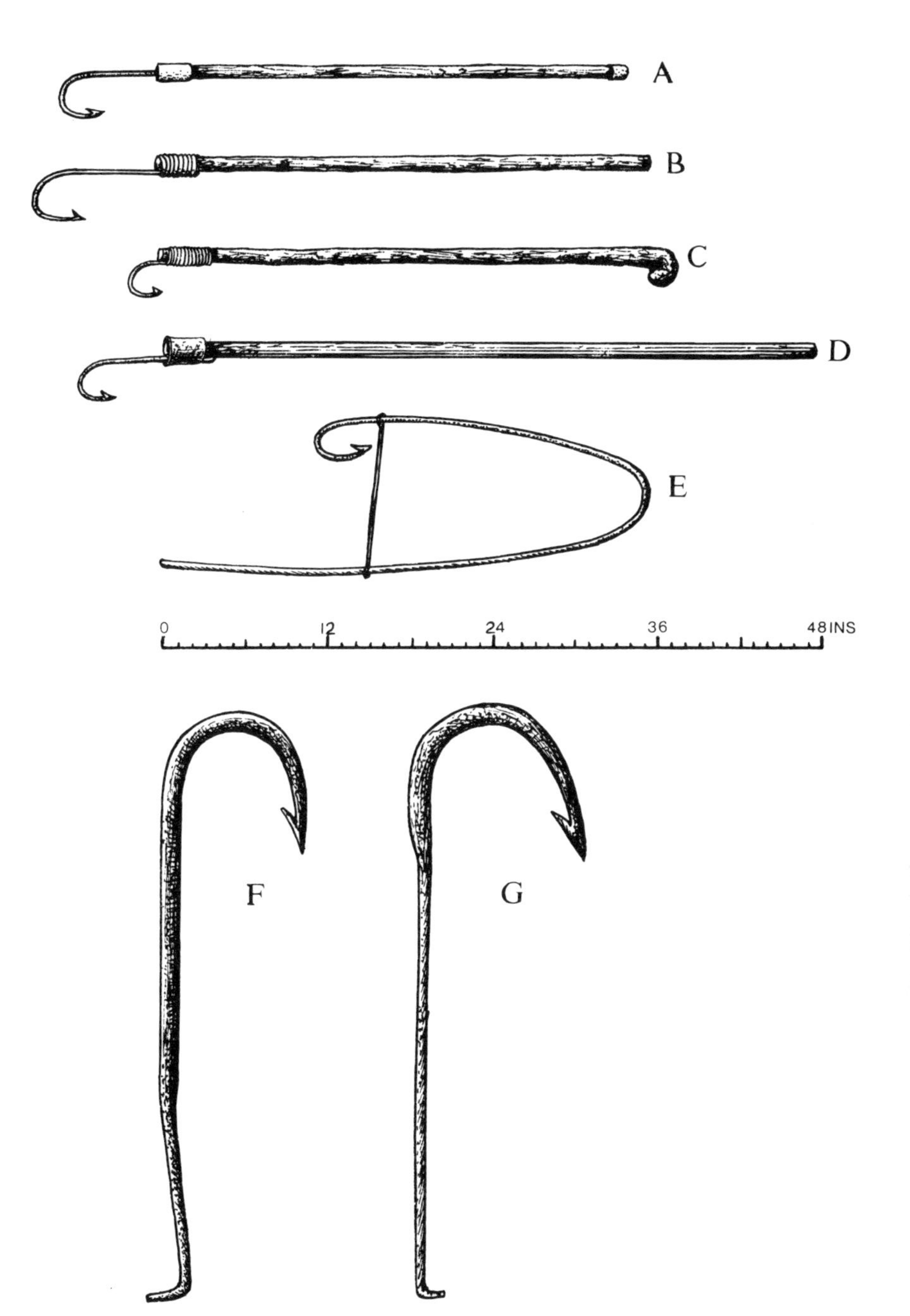

Salmon gaffs: A Carmarthenshire; B Cardiganshire; C Cardiganshire; D Pembrokeshire; E Cardiganshire; F Carmarthenshire; G Cardiganshire

to west Carmarthenshire said that a torch *(ffagl)* was prepared by placing two or three sacks in an oven to dry thoroughly. The sacks were then attached firmly to a stick and soaked with paraffin. A junior member of the poaching team would be given the task of carrying the paraffin 'jack' to the riverside, and it was his responsibility to pour more paraffin on the sacking at regular intervals to ensure that a steady light was provided to attract the salmon, which was then gaffed as near the head as possible. In textile-producing districts, such as the Teifi Valley, the waste from wool carding machines, that had already been soaked in oil during the carding process, was in considerable demand from the poachers. An ordinary farm pitchfork was thrust into a heap of this and set alight. Alternatively, a metal cage possibly made of old fencing or cask hoops could be improvised in such a way as to make a receptacle for burning fuel.

By the nineteen-thirties, burning the water with a lighted *ffagl* had virtually disappeared and acetylene or carbide cycle lamps were regarded as essential parts of poaching equipment. Oil pressure lamps, such as the Tilley and Aladdin, that were to be found as outdoor lamps on many Welsh farms, were regarded as being particularly suitable for salmon poaching. A piece of tin was usually wrapped around the lamp, so as to concentrate the light in one particular direction and to conceal its beams from prying eyes. In more recent years electric torches have replaced paraffin lamps and a car fog lamp, used in conjunction with a radio dry battery, is regarded as an excellent though wasteful source of power. The two wires connected to the lamp have to be changed from cell to cell on the battery at frequent intervals, as soon as the light from one pair begins to dim.

Just as single-barbed gaffs are universal poaching instruments, so too are multi-bladed salmon spears.[9] Although salmon spearing was forbidden as early as 1532[10] by Act of Parliament, a restriction that was reinforced and reiterated at frequent intervals ever since, fish spearing has continued as a commonplace practice to the present day.[11]

A salmon spear differs from an eel spear in that it is designed for poaching the fish with a barbed tine or pointed tine rather than in forcing the fish in between flattened tines as on eel spears. The number of tines on a salmon spear may vary from three to as many as nine, and the size of spear will depend on the type of fish expected in a river at any one time and the

A West Wales poacher with salmon spear

preference of the individual fisherman. Thus for example, one informant in West Wales said that his poaching activities began in October with the capture of migratory sewin *(twps y dail).* For this purpose he used a three-tined spear, each tine barbed, but measuring no more than 5 inches wide and 6 inches long. Later, with large salmon in the river he used a heavy seven-tined spear, 12 inches wide and 12 inches long. In West Wales another informant said that poaching would begin after Halloween, and would continue until January or February, or 'until one was fed up with salmon'.

Salmon spears vary tremendously in shape and design and most would be made by local blacksmiths. In some cases the tines would be pointed and notched. All were equipped with sockets for the insertion of a wooden handle, and many poachers were of the opinion that the longer the handle, the more efficient the spear. Unlike fishermen in Scotland, Welsh salmon poachers never used a handle-less casting spear.[12] In some cases if a salmon were seen in a particularly deep pool, a rope would be attached to the handle and the spear thrown into the pool with considerable force, the fisherman holding on to the rope. Great skill was necessary to use a spear in this way.

When poaching was widely practised as a rural pursuit, there was considerable co-operation between neighbouring farmers. In West Wales one farmer would supply a spear and another would be expected to supply a light. The spoils were then divided equally between the parties involved. D. J. Williams in his classic of Carmarthenshire life[13] gives a vivid account of a poaching party: 'Often towards the end of October and the beginning of November silent walking would be in progress along the banks of the river to ascertain where the salmon were burying. Once annually towards this time, three or four places each side of the vale . . . would devote a whole day to taking salmon and that . . . in broad daylight, not in the middle of the night with the help of the dazing torch, as was generally done . . . The spoils were fairly shared at the end of the day.' He describes how one person used 'a pitchfork or a fork straight from the cowhouse . . . not the barbed hooks or the heavy

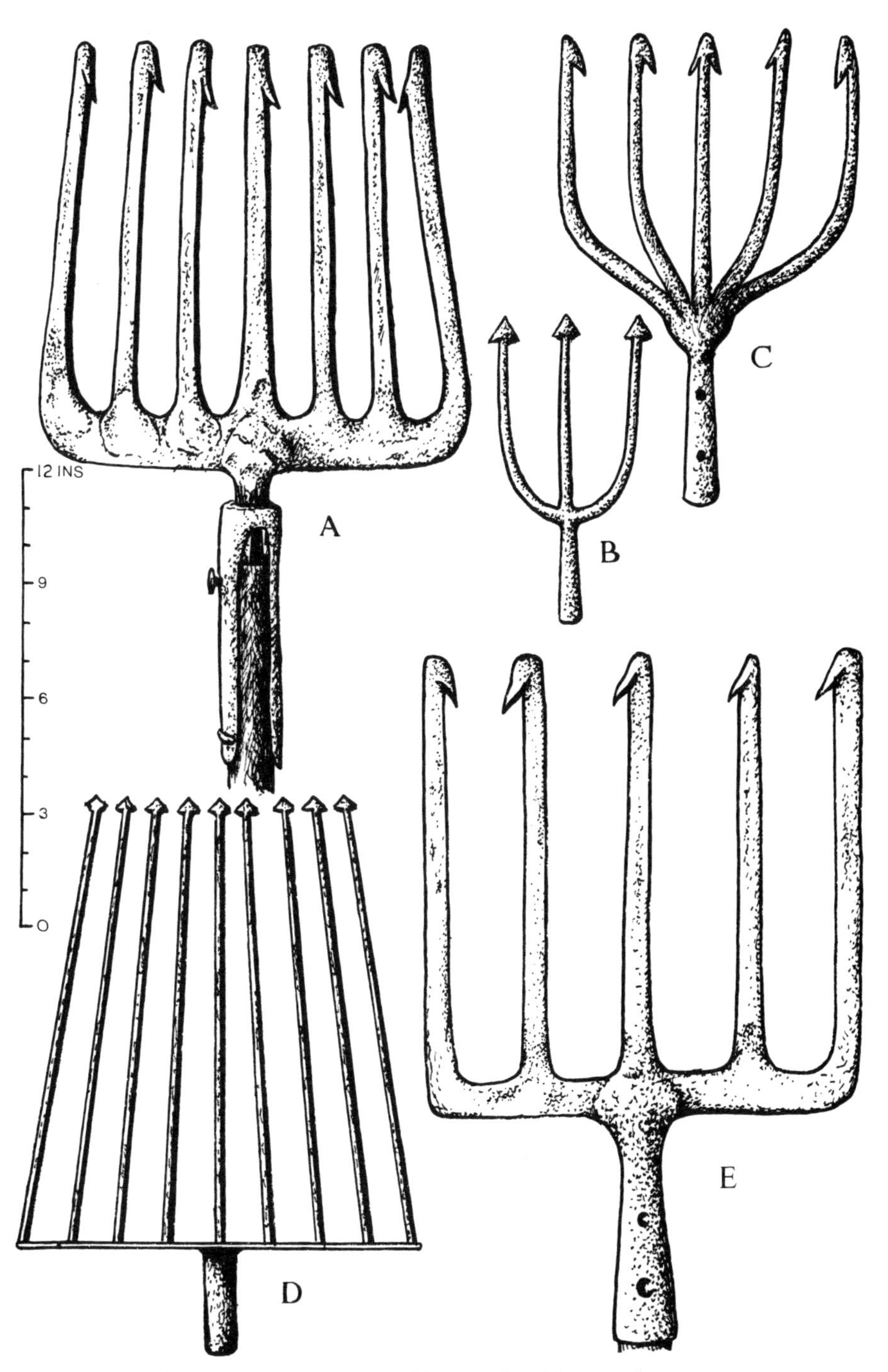

Salmon spears: A Radnorshire; B Flintshire; C Glamorgan; D Carmarthenshire; E Pembrokeshire

gaff made by William the Blacksmith'.

Before venturing on a poaching expedition it was customary to file the tines of a spear to a fine point, and on arriving at a pool with the light in position, spearing began. As soon as a salmon appears and is attracted by the lights, the spear is thrust down as near the head of the fish as possible. The person who spears a salmon along its body is regarded as a very poor fisherman, for it is important that the edible flesh is not damaged in any way.

Quite often in rural Wales a salmon spear is by custom passed on from father to son, and many informants in West Wales when interviewed mentioned spears that had been passed on from father to son over three or four generations.

As an alternative to the spear and gaff a snare or wire loop mounted on a stick may be used, the noose being slipped over the fish. The snare is the forerunner of the angler's 'tailer' which is only legal if used as an auxiliary to a rod and line.

A poaching implement that is found in all parts of Wales, but especially in the rivers of north-east Wales, is the so-called 'snatch', 'stroke haul' or 'Aberaeron minnow'. This consists of a hook or hooks suspended on the end of a line, attached to a

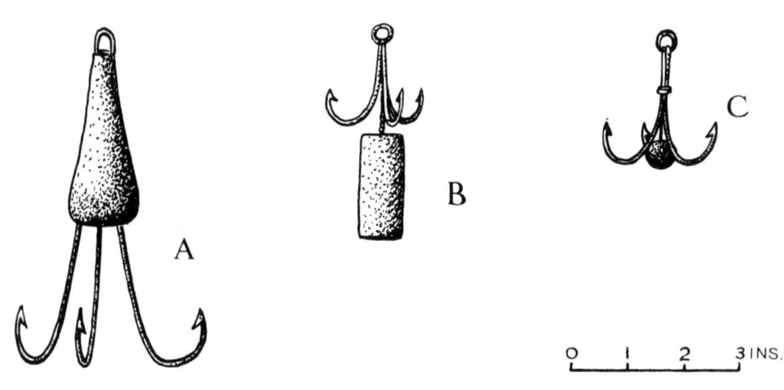

Aberaeron minnows: A Wye; B Usk; C Dee

piece of string and usually a rod. This is cast among the shoal of fish and then suddenly jerked, and this foul hooks the fish. This method of fishing is particularly popular amongst shoals of sewin congregating below weirs and waterfalls. The size and weight of a 'Aberaeron minnow' can vary considerably. Three examples in the Welsh Folk Museum Collection, all confiscated from poachers on the Dee, are all triple hooks, 6 inches, 4 inches and 3.5 inches long respectively. A lead weight is fixed to the top of two of the minnows, while in the other, the weight hangs down below the three hooks.

Another illegal instrument that uses fishing hooks is the 'lath otter board' *(dwrgi)* used on lakes and the larger rivers. This consists of a wooden board weighted at one edge to make it float upright, to which is attached a line, so arranged that when it is pulled, the angle between the otter and the line causes the former to move parallel to the shore. Onto the line is fastened a number of short leaders with artificial flies or lures attached.

In describing an otter from the Tal-y-llyn district of Merioneth a correspondent[14] says, 'The otter *(styllen)* was usually used on lakes, after dark between the months of May and September. The implement being used was the *styllen:* a plank of wood 14 inches long by 10 inches wide with a strip of lead on one side, so that it floated upright. There were two holes bored in its centre, and through these holes a piece of cord was tied in a knot visible on both sides. The ends of the board were pointed. The next piece of equipment required was a fly carrier *(car plu)*, a slab of thin planking with about 40 yards of line attached to it. They were equipped with large flies, each 15 inches from the next. The fishing line was attached to the cord and the *styllen* and this was launched. Instead of following the shore, the otter would travel towards the centre of the lake and would return to the bank according to the length of fishing line. I can assure you that it was (and is) a most profitable method of fishing.' The otter was made illegal in 1861 but its use was widespread in all parts of Britain until recently.[15] They varied in size and shape, but all were equipped 'with flies dangled at the ends of short lengths of gut as "droppers", about 6 inches long, the upper

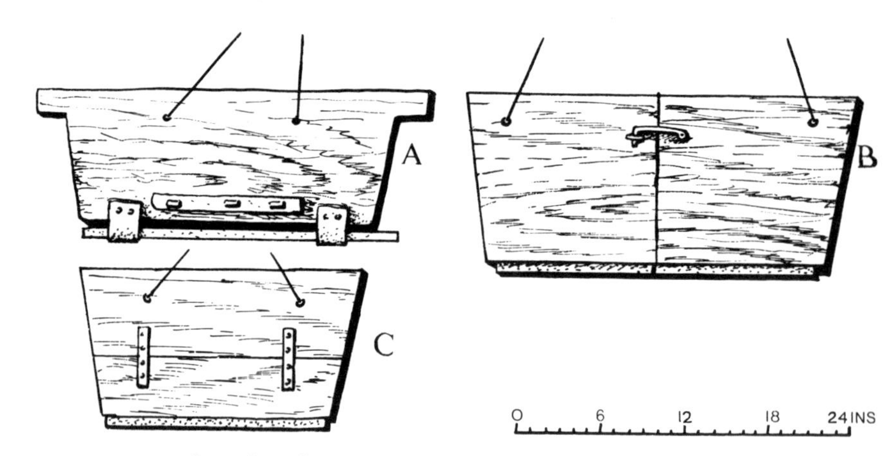

Otter boards: A Cardiganshire (Llyn Eiddwen); B Usk C Merioneth

ends of which were tied to the main line. Sometimes the droppers were fastened to the little metal swivels spaced at intervals along the line, and kept from moving more than an inch or two by a knot on either side of each swivel or by two little balls of *guttar percha* about the size of peas'.[16] In some cases the gut and flies could be attached directly to the otter board, with four or six on each one. A piece of cord was attached to this for drawing it back to the shore, but in some cases the otter was allowed to drift with the current and was not under the control of the fisherman in any way; their weighting and the movement of water was sufficient to ensure their safe return to the bank. Usually otters were left hidden on the banks of a lake or river and very rarely carried home.

In recent years, poachers have moved away from the banks of rivers and brooks and can now operate under water using modern underwater breathing equipment and harpoon guns. In these guns strong elastic or compressed air is used as a propellant. This method of poaching is a long way removed from simple 'groping' or 'tickling', that consists of nothing more than the use of bare hands to locate and grasp or lure fish, under the roots of a tree or beneath a stone in the river. Despite the skill, patience and artistry involved in this simple method of fish capture, it is still illegal. The above methods of poaching are only capable of taking one, or, in the case of the otter board, a few fish at a time, and for this reason, the damage caused by one instrument is relatively limited. Far more damage is done by methods of mass destruction, such as nets, explosives and poisons.

Of the illegal nets the most commonplace are the small-meshed gill nets to entangle the fish. These, which were described in Chapter 7, are set across the flow of a river and sunk below the surface of the water. For this reason they are very difficult to detect especially in deep, narrow sections of river, such as the notorious Cilgerran gorge on the Teifi. The nets may be simple gill nets, up to 30 yards in length, or they may be complex fixed trammels. Stake nets, set between high and low tide marks in an estuary or along the sea shore, are also illegal instruments more often than not, as are seine nets of illegal mesh or used by unlicensed fishermen in many estuarine waters. They may be ordinary short seines used in combination with a boat or they may be wade nets. Scoop nets of fine mesh, such as the now defunct perk nets of the Wye, which are related to the lave net in type, may also be used for the illegal capture of salmon at falls and weirs.

Electro-fishing is a recognised fishing management technique for such tasks as controlling fish stocks, but electrical equipment can only be used legally with the permission of the river authorities. It is nevertheless practised by poachers using possibly a car battery or portable electric motor. This technique passes an electric current through the water, and either stuns the fish in the immediate vicinity of the electrodes or drives them into a net.

The use of explosives, by placing a small charge in a pool containing fish, will kill everything within the critical range, irrespective of size. This is a most wasteful method of poaching as only a small proportion of the fish killed are of value, but the most detestable of all poaching methods involves the introduction of toxic substances into a stream. Often the operators do not realise the potency of some of the substances they use and they invariably kill large numbers of fish over long lengths of water, yet perhaps only have the facilities for taking perhaps a dozen or so.

By tradition the poacher has been regarded as an heroic figure, but no one knows with any degree of accuracy what the effect of poaching on fish stocks is likely to be. 'If one area', says Millichamp, 'has only ten expert poachers in a spawning season of ten weeks, they need average only one fish per day each for that area to lose seven hundred potential spawning fish, mostly females.' Under the *Salmon and Freshwater Fisheries Act of 1972*, penalties of up to £400 and two years in prison or both may be imposed by a Crown Court for poaching. 'Generally', adds Millichamp, 'much smaller penalties in the form of nominal fines are imposed when cases are heard in local Magistrates' Courts . . . A poacher fined £10 for taking salmon illegally laughs it off and goes out that night to make enough from poaching to pay the fine and also to make a handsome profit!'

For notes to Chapter 11, see pages 324-5

Notes

Chapter 1: Introduction (pages 13-30)

1 Giraldus de Barri: *The Itinerary of Archbishop Baldwin through Wales in 1188* [1806 ed.], pp 36, 48.
2 Owen, George: *The Description of Pembrokeshire* [1892 ed], Vol 1, p 117.
3 Donovan, E.: *Descriptive Excursions through South Wales* [1805], Vol 1, p 237.
4 Pennant, T.: *British Zoology* [1776], Vol III, pp 336-7.
5 For details of customs associated with coastal communities, see Jenkins, J. G.: 'The Customs of Welsh Fishermen', *Folklore*, Vol 83 [1972], pp 1-19 and Jenkins, J. G.: 'Llangrannog – Some Aspects in the Development of a Coastal Village', *Ceredigion*, Vol III [1959], pp 231-45.
6 *Report of the Commissioners on Salmon Fisheries* [1861], Introduction to Minutes of Evidence, p iv.
7 The general name for a salmon in Welsh is *eog*. *Gleisiad* is also used, but there is considerable confusion relating to salmon at different stages of growth. A correspondent in *Bye-Gones* (August, 1878), p 73, gives the following terms: i. *Silod y gro, silod y gôg*, ii. *Silod brithion, brith y gro*, pl *brithroaid*, iii. *Brithyll y môr, brithyll brych*, iv. *Gwyniad y gôg*, v. *Gwyniad hâf*, vi. *Gaflaw*, vii. *Adfwlch*, viii. *Maran*, ix. *Camog*, x. *Chwiwell*, xi. *Gleisiad*, xii. *Penllwyd*. Few of these terms are used by fishermen today.
8 Known as *peal* in Devon, *sprod* in Cumberland and *white trout* in Ireland. In Welsh the sewin or sea trout is known as a *gwyniedyn* [pl *gwyniad*], *penllwyd, eog frithyll, mwlyn* and *gaflaw*. In West Wales the large autumn sewin is known as *brithyll y dail* (trout of the leaves) or *twps y dail* (idiots of the leaves).
9 Including Dysynni, Mawddach, Artro, Dwyryd and Glaslyn.
10 Figure for 1950. 1949 figure not available.
11 Including English section of river.
12 Based on figures in *Report of Committee on Salmon and Freshwater Fisheries* (HMSO, 1961).

13 *Salmon and Freshwater Fisheries Report* (1961), *op. cit.*, p 3.
14 *Ibid.*, p 4.
15 *Report of the Commissioners appointed to inquire into Salmon Fisheries* (HMSO, 1861).
16 *Op. cit.*, p 4.

Chapter 2: Fishing Weirs and Stake Nets (pages 31-43)

1 Professor Melville Richards of the University College of North Wales, Bangor, has compiled a list of fishing terms in Welsh place names, to be published in *Folk Life* [1974].
2 *Argae,* pl *argaeau,* is derived from the prefix *ar* + *cae* (enclosure); *cored,* pl *coredau,* is derived from *cor* (plaiting or binding) + suffix *ed. Cf.* Old Breton – *coret;* Irish – *cora.*
3 Davis, F. M.: *An Account of the Fishing Gear of England and Wales* [London, HMSO 4th edition, 1958], p 1.
4 *Report of the Commissioners appointed to inquire into Salmon Fisheries (England and Wales)* [London, HMSO, 1861], pp ix-xi.
5 *Ibid.*, p xxiv.
6 28 & 29 Vict. c. 121.
7 13 & 14 Geo 5, c. 16.
8 Williams, D. H.: *The Welsh Cistercians* (Pontypool, 1970), p 75.
9 *Ibid.*, p 76.
10 Owen, G.: *The Description of Pembrokeshire* (1603) (London, 1892 ed).
11 Jones, Thomas (ed): *Gerallt Gymro* (Cardiff, 1938), p 116.
12 Phillips, J. R.: *The History of Cilgerran* (London, 1867), p 180.
13 *Ibid.*, p 181
14 Williams, David: *The Rebecca Riots* (Cardiff, 1955), p 225.
15 Davies, H. R.: *The Conway and the Menai Ferries* (Cardiff, 1942), p 108.
16 Davis, *op. cit.*, p 1.
17 Minutes of Evidence, *op. cit.*, p 152.
18 Lewis, E.: 'The Goredi near Llanddewi Aberarth', *Arch. Camb.*, Series 7, Vol IV, [1924], pp 295-8.
19 *Ibid.*, p 395.
20 In *Hanes Taliesin* of the 6th century AD. *Gored Wyddno* is mentioned 'Ac yn yr amser honno yr Gored Wyddno yn y traeth rhwng Dyfi ac Aberystwyth'. [At that time Gorod Wyddno was on the beach between the Dyfi and Aberystwyth.]
21 Davies, D. C.: 'The Fisheries of Wales', *Liverpool National Eisteddfod Transactions 1884* [Liverpool, 1885], p 309.
22 Matheson, C.: *Wales and the Sea Fisheries* [Cardiff, 1929], p 63.

23 Davies, H. R., *op. cit.*, p 282.
24 Evans, J.: *Letters Written during a tour through North Wales in the year 1798 and at other times*[London, 1804], p 231.
25 Senogles, D.: *Ynys Gorad Goch* [Menai Bridge, 1969].
26 *Ibid.*, pp 95-6
27 *Report of Committee on Salmon and Freshwater Fisheries* [London, HMSO, 1961], p 146.
28 *Minutes of Evidence, op. cit.*, p 205.
29 A salmon coop on the River Derwent in Cumbria is still in regular use, a licence fee of £100 per annum being payable to the Cumberland River Authority. In 1970 369 salmon weighing a total of 2,469lb was caught in this trap. The record season within the last twenty years was that of 1963 when a total of 1,351 salmon weighing 8,312lb was caught. The licence fee for the Caerhun weir is £30 per annum payable to the Gwynedd River Authority.
30 Davis, *op. cit.*, p 30. The salmon garth at Ravenglass on the River Esk in Cumbria, licensed at £50 per annum, seems to be the same type as that used in Swansea Bay. This was described by Davis (p 29) as follows: 'The "Salmon Garth" at Ravenglass is interesting, as it is made partially of net, and partially of hedging or "weiring". The dimensions of the inner circle or "house" being about 10 x 8 feet, with entrance similar to that found in the cage of the "hedge baulks", but without a step. The stakes project 9 feet above the ground, and are about 6 feet apart; between the stakes is a wall of weiring about 3 feet high. In the "house" the weiring is only about 6 inches high. The stakes and weiring stand all the year round, but the net is only mounted during the season, and is strung on a head-line and foot-rope between the top of the stakes and the top of the weiring. The mesh is of the size allowed by the local regulations (2 inch bar).'
31 *Ibid.*, p 35.
32 *Second Annual Report, op. cit.*, p 15.
33 Matheson, C.: *op. cit.*, p 65. The 'butts' described by Matheson were possibly of woven withy, similar to the 'putchers' and 'eel butts' that will be described later.
34 *Minutes of Evidence, op. cit.*, p 193.
35 *Ibid.*, p 210
36 *Report of the Commissioners, op. cit.* [HMSO, 1861], p xi.

Chapter 3: Putchers, Putts and Basket Traps (pages 44-66)

1 See Chapter 2.
2 24 and 25 Victoria Ch. 8: 28 and 29 Victoria Ch. 121: 13 and 14

George, Ch 16.

3 John, D.: *Fly Fishing on the Usk* [Brecon 1968], p 16, quoting from the *Report of the Clerk of the Usk Conservators* (1902).

4 Davis, F. M., *op. cit.*, p 49.

5 *Op. cit.*, p 5.

6 Grimble, S. A.: *The Salmon Rivers of England and Wales* [London (c. 1905)], p 88.

7 *Minutes of Evidence, op. cit.*, p 11.

8 John, *op. cit.*, p 16.

9 *Second Annual Report on Salmon Fisheries* (1863), p 26.

10 These are deposited at the Gloucester County Record Office.

11 *Report of Committee on Salmon and Freshwater Fisheries* (HMSO 1964), p 14.

12 Messrs. Proctor Bros., Wire Works, Caerffili, Glamorgan.

13 *Survey of the Manor of Goldcliff.* Monmouthshire County Record Office, Newport, MSS 258.

14 Matheson, *op. cit.*, p 66.

15 Mr Wyndham Howells, to whom I am grateful for a great deal of information on fishing in the Goldcliff district.

16 *Survey of the Manor of Goldcliff, op. cit.*

17 I am grateful to Mr Wyndham Howells for this information.

18 *First Annual Report, op. cit.*, p 5.

19 Jenkins, J. Geraint: *Traditional Country Craftsmen* (London 1965), pp 37-47.

20 Wilshire, L.: 'Salmon Fishing in the Severn Estuary', in *Country Life*, , 27 February 1952, p 576.

21 Waters, B.: *Severn Tide* [London 1947], p 152.

22 *Ibid.*, p 572.

23 Wilshire, *op. cit.*, p 576.

24 Transcript of Appeal. John Jones and the Justices of the Peace for the Division of Nant Conway, Caernarvonshire Amateur Session, 20 October 1887. (In possession of Mr Owen Goodwin of Tan'rallt, Betws-y-coed, the present licensee of the fish trap. I am grateful to Mr Goodwin for a great deal of help in tracing the history of fishing on the Lledr.)

25 Grimble, *op. cit.*, p 187.

26 Transcript of Appeal, *op. cit.*

27 Under Section 11 of the Fisheries Act of 1861.

28 *Liverpool Daily Post*, 31 September 1924.

29 13 and 14 Geo. V CL 16, p 27.

Chapter 4: Net-Making (pages 67-80)

1 Brayley, E. W. & Britton, J.: *Beauties of England and Wales* (London, 1801), Vol. IV, p 519.
2 Claridge, J.: *General View of the Agriculture . . . of Dorset* (London, 1793), p 26.
3 *The Sunday Times*, 7 November 1971, p 67.
4 Local synonyms – shale; chale; pin; moot; cowl; keevil; kibble and mesh pin.
5 Davis, F. M.: *An Account of the Fishing Gear of England and Wales* [London, HMSO, Fourth Impression, 1958], p 7.
6 Local synonyms – letting out; rising; stealing; making; sped; hitching; widening; putting in; half knees.
7 Local synonyms – taking in; stealing; shrinking; narrowing.
8 Davis, *op. cit.*, p 9.
9 Local synonyms – hanging meshes; 'square meshes'.
10 Davis, *op. cit.*, p 10.
11 Local synonyms: back rope; top rope; cork line; upper tant; gale; head-rope; top bank; flue.
12 Local synonyms: back rope; sole rope; ground rope; lead line; bottom line; lower tant; gale.
13 Davis, *op. cit.*, p 15.
14 *Ibid.*, p 15.
15 *Ibid.*, pp 15-16.
16 James Sallis, St Dogmaels.
17 Davis, *op. cit.*, p 22.
18 I.E.B.C. (editor): *Facts and Useful Hints relating to Fishing and Shooting* [London, 1874], p 118.
19 *Ibid.*, p 119.

Chapter 5: Push Nets (pages 81-92)

1 Davis, F. M.: *An Account of the Fishing Gear of England and Wales* (HMSO, 1937 ed), pp 121-2, describes the principal push nets as follows: 'i. Southport "Power net". Beam ("power head"), 6 feet. Staff 8 feet. The upper boundary of the mouth is formed by the two "bows" or "benders" which overlap for about 18 inches of their length. This distance from the beam to the bow is about 2 feet 6 inches. The net is as follows: Round netted. Tail 70/70, 2 feet 6 inches. Then braid 70/320, creasing twice every round. Then leave 70 meshes bosom, and continue backwards and forwards on the rest, bating once every round on the selvage, and finish off with about 2 feet 6 inches of straight net. This forms a shape corresponding to the square and wings of a trawl upside down. Mesh usual. Material, No 12 cotton. Dressing, boiled oil. In some cases this type of net is

known as a "Shoe net" (eg *Cumberland*). ii. "Pandle" of the South Coast. Beam, 10 feet. Staff (driver), 7 feet. Upper beam ("spread net"), 10 feet. The distance between beam and upper beam about 2 feet. The net is simply a flat piece about 10 x 4 feet, the extra depth allowing of a bag to be formed. iii. Lytham "square net". Collapsible net of the pandle type, but with a specially made bag and cod-end. Beams 7 feet. Stanchions, 1 foot 6 inches. iv. Cleethorpes "Shove net". Beam, 8 feet. Staff, 8 feet. Side irons, 3 feet. There is a special ring apparatus for tightening and stretching the net. v. Somerset "skimming net". Poles, 10 feet, hinged for folding and kept in position by a short cross bar, the "spreader". Pushed along by a man standing in the small angle. Largely used for flounders.'

2 Waters, Ivor: *About Chepstow* (Chepstow, 1952), p 35.

3 Information from Mr M. Thomas, Ferryside [interviewed November, 1969] and Mr E. Brown, Laugharne [interviewed August 1971]. The South West Wales River Authority allow the licensing of 'Heave or Lamp nets' at £5 per annum, although no licences have been issued in recent years.

4 *Usk River Authority: Fishery Bye-laws* (1963), p 10.

5 *Wye River Authority: Fishery Bye-laws* (1953), p 6.

6 Taylor, J. N.: *Illustrated Guide to the Severn Fishery Collection* (City of Gloucester Museum, 1953), p 13, quoting from Smyth, John: *Hundred of Berkeley* (Berkeley MS, ed Sir John Maclean) (1885), Vol III, p 321.

7 *Minutes of Evidence, op. cit.*, p 49.

8 The shadd is a close relative of the herring and is a very bony fish.

9 Welsh Folk Museum Accession Number, 71.44/3.

10 Taylor, *op. cit.*, p 13.

11 Wilshire, L.: 'Salmon Fishing in the Severn Estuary', *Country Life*, 27 February 1953, p 577.

12 Davis, *op. cit.*, p 123.

13 *Second Annual Report, op. cit.*, p 33.

14 'Commercial Salmon Fishing in the River Parrett', *The River Boards Association Year Book* (1953), p 15.

15 (1960 ed), para 15 g.

16 Information in a letter from F. Ackroyd, Fisheries and Pollution Officer, Lincolnshire River Authority, 12 April 1972.

17 Information in a letter from M. Ll. Parry, Fisheries Officer, Yorkshire River Authority.

18 Johnson, R. S.: *The River Boards Association Year Book* (London, 1953), pp 32-3

19 Sanderson, S. F.: 'Haaf Net or Have Net: A Contribution to the

Study of Northern Cultural Connections', *Report of the Fifth Viking Conference in Torshaun* (July, 1965), p 134.

20 *Cumberland River Authority Bye-laws* (1972 ed), p 4.

21 *Second Annual Report, op. cit.*, p 34.

22 Johnson, *op. cit.*, p 32.

23 Sanderson, *op. cit.*, p 135.

24 Kissling, W.: 'Tidal Nets of the Solway', *Scottish Studies,* Vol 2, No 2 (1958), p 170.

25 *First Annual Report of the Inspectors of Salmon Fisheries (England and Wales)* (London 1862), p 16.

26 Sanderson, *op. cit.*, p 134. In this paper Sanderson gives a possible derivation of the word 'haaf'. He says: 'The problems of provenance and dating are touched on briefly by Kissling in his survey article on tidal nets. Recalling that the Norse word *haf = the open sea* survives on a loan word in Hebridean Gaelic and in the dialect of Shetland he states: "It is at least a plausible guess that the net came over to this country with the Norsemen, but there is no evidence". I wish to suggest an alternative theory which, while still a guess, seems to me more plausible. The theory is founded on any exhaustive enquiry into the lexical evidence and the evidence of the objects to which the words relate; and the kernel of the theory is to be found in a sentence added, after friendly discussion, to Kissling's article in *Scottish Studies* at my editorial suggestion. This sentence reads "Another explanation of the name links it with Icelandic *hafr,* a poke-net for herring fishing, Norwegian *hav,* connected with the verb *hefja* meaning 'to lift or raise'".

In examining the lexical evidence one has various glossaries to consult as well as standard dictionaries of the calibre of the *English Dialect Dictionary* and the *Scottish National Dictionary.*

The quality of these sources varies considerably: and it is worth noting, as a typical example of how much help can, and how much cannot, be extracted, from such sources by the ethnographer, that only one out of the three relevant glossaries in the English Dialect Society series cites the word haaf-net, and this one gives a rather vague and unhelpful definition and comments inadequately on distribution. The work in question is Ellwood's *Lakeland and Iceland: A Glossary of Cumberland, Westmorland and North Lancashire* (1895, p 150), which cites the verb *haaf:* 'to fish with the large *haaf* or sea-nets. Icel. *haf'r* and *hafnet,* a net with a poke-formed centre to collect fish in. This word is used by fishermen of the Solway, both on the Scottish and Cumbrian side'. But the spoon-net or dip-net, a small poke-net hanging from a circular frame and equipped with a single horizontal handle, is called in Norwegian *hav,* Swedish *hav,*

Danish *hov.* A similar net with a much longer handle, used for trapping birds in Iceland, is known by the same name, *hafur.* This is not the same as the Faroese *fleygastong:* the frame is a complete wooden circle. The word *hav,* like *hav* = sea, has also been borrowed into Scottish Gaelic as *tabh* with the meaning *spoon-net:* an illustration is to be found in Dwelly's dictionary. For reasons of linguistic history it can be said that this borrowing is unlikely to have taken place before the 12th century. The word is also listed in Marwick's *Orkney Norn* in the form *heevie,* with two meanings, firstly 'a creel for bait', and secondly 'a small spoon-net'. Before leaving the linguistic material, note should be taken of the French word *haveneau* or *havenot:* Littré defines this as follows: Nom d'un petit filet formant une espèce de poche conique, tenue ouverte par un cercle sur lequel il est transfilé; un manche assez léger sert à le diriger pour pêcher des crevettes . . . This is, once again, a small spoon-net.

Here then there would seem to be a better etymology for the word *haaf-net.* It is not a deep-sea net, but a poke-net, which is raised from the water when the fish enter it (hafr is connected with *hefja = to raise*). The form of the haaf-net, however, appears to be unique: this particular shape of frame is not reported from the Scandinavian countries or elsewhere. It is important to observe, also, that the haaf-net is found in the British Isles only in an area of Norse colonisation and settlement.

Nodal and Milner's *Lancashire Glossary* and Dickinson's *Cumberland Dialect* both ignore the word, but the later Cumberland glossary of Dickinson and Prevost (not in the English Dialect Society series) defines *haaf-net* as 'a pock-net fixed on a frame of wood . . .' This is also cited, with a reference to Provost, in Brilioth's *Grammar of the Dialect of Lorton.*

The English Dialect Dictionary cites *haaf-net* with references from Dumfriesshire, Wigton and Cumberland, but not Lancashire: none of the descriptions of the form of the net would be clear to anyone who had not actually seen the net himself, but undoubtedly they do refer to the net under discussion here. The word is etymologised from *haf = sea.*

Ellwood's definition of the verb *haaf = to fish with the large haaf or sea-nets* and Brilioth's *af-net = a pock-net, a sea-net* bring us to the heart of the problem. In the dialects of Shetland, Orkney and the Hebrides, the word *haf* always means *the deep sea.* The philologist in his library may be content to etymologise *haaf* as *sea* and define *haafing* as *fishing with the large haaf or sea-nets* (the word *large* is incidentally a meaningless adjective in this context), but he has only to go to the river bank and watch a haaf-net fisher at work to know that this is not a deep-sea net, and common-sense should prompt him to seek another explanation from his reference books.

The *Scottish National Dictionary* has several references under *halve-net* and etymologises from O. N. *hafr* which it glosses as *a similar kind of net.* This is a more likely derivation, but the actual definition begs a number of questions. There is, typogically speaking, no similar kind of net; the haaf-net, in the form described here, does not appear to be known in the Scandinavian countries.

Chapter 6: Stop-Net Fishing (pages 93-107)

1 *Minutes of Evidence taken before the Commissioners appointed to inquire into Salmon Fisheries (England and Wales)* (London, 1861), p 126.
2 Large, N. F.: 'Stopping boats of the Severn', *Country Quest* (November 1969), p 3.
3 Taylor, J. N.: *Illustrated Guide to the Severn Fishery Collection* (Gloucester 1953), p 14.
4 Large, *op. cit.*, p 2.
5 *Certificates of Privileged Engines* (1863-1868). At the Gloucestershire County Record Office.
6 Large, *op. cit.*, p 4.
7 *Report of the Commissioners, op. cit.*, p 10.
8 Waters, Ivor: *About Chepstow* (Chepstow 1952), pp 33-9. The fish house at Chepstow is still known as 'Stuart House' since Alexander Miller and his brother David sent salmon to London to a firm called Stuart.
9 *Wye River Authority Fisheries Department Annual Report* (1969), p 5.
10 *Ibid.*, p 5.
11 *Usk Board of Conservators* (Newport 1936), p 13.
12 *Minutes of Evidence, op. cit.*, p 127.
13 Messrs Harold Jenkins; Garfield Richards; J. Griffiths and William Phelps of Hook, and Messrs Llewellyn; Glyn Morgan; Reg Jones and B. Jones of Llangwm.
14 An example of a Llangwm boat is preserved at the Welsh Folk Museum, St Fagans. This was built by John Palmer, a Llangwm fisherman, in 1890.
15 Nicholls, F. F.: 'Compass Netsmen of Llangwm', *The Countryman* (Summer 1961), p 264.
16 *Ibid.*, p 264.
17 *Ibid.*, p 266.

Chapter 7: Coracle Fishing (pages 108-99)

1 Malkin, B. H.: *The Scenery, Antiquities and Biography of South Wales* (1807), Vol II, p 206.
2 *De Bello Civile,* Book 1, Chap IV.
3 *Naturalis Historie,* Book IV, Chap 30 and Book VII, Chap 27.
4 For example, Caius Julius Solinus, Avierus and Sidonius Apollinaris all make references to skin-covered boats of the sea-going variety.
5 Evans, J. Gwenogvryn (ed): *White Book of Mabinogion* (1907), p 47, 'par weithon wahard y llongeu ar ysgraffeu ar corygeu ual nat el neb y gymry'. The coracles were obviously designed for undertaking the long sea-voyage from Ireland to Wales.
6 Hornell, J.: *British Coracles and Irish Curraghs* (1938), pp 74 et seq.
7 Williams, Ifor (ed): *Canu Aneirin* (1938), p 44. Sir Ifor Williams suggests that the word 'coracle' is derived from the Latin 'corium'.
8 Williams, S. J. and Powell, J. E.: *Cyfreithiau Hywel Dda yn ol Llyfr Blegywryd* (1942).
9 Jones, T. (ed): *Gerallt Gymro* (1938), p 203. This translation is more accurate than that of the 1806 ed of Giraldus de Barri, *The Itinerary of Archbishop Baldwin through Wales, ADMCLXXXVIII,* p 332.
10 In the earlier translation the coracles of 'twigs . . . covered both within and without with raw hides' gives a false impression of their construction.
11 Owen, George: *Description of Pembrokeshire* (1892 ed), pp 118-19.
12 *Royal Commission to Inquire into Salmon Fisheries* (1863), p 141.
13 Cardiff Public Library. A translation of the *cywyddau* appears in Lord Mostyn and T. A. Glenn: *History of the Family of Mostyn of Mostyn* (1925), pp 35-41.
14 Hawkins, J.: Izaak Walton's *Compleat Angler* (1760 ed), footnote, p 33.
15 Parry, J. Hughes: *A Salmon Fisherman's Notebook* (2nd ed, 1955), pp 18-19.
16 *Report of the Commissioners appointed to Inquire into the Salmon Fisheries (England and Wales)* [1861], p 224.
17 Hawkins, *op. cit.,* p 33.
18 Lloyd, J. E. (ed): *History of Carmarthenshire* (1939), Vol II, pp 319-20.
19 Donovan, E.: *Descriptive Excursions through South Wales and Monmouthshire* (1805), Vol II, p 229.
20 Lloyd, *op. cit.,* II, p 320.
21 Donovan, *op. cit.,* p 228.

22 Hornell, *op. cit.*, p 16.
23 Phillips, J. R.: *History of Cilgerran* (London 1867), p 167.
24 F. C. Llewelyn, Cenarth.
25 Jenkins, J. G.: *The Welsh Woollen Industry* (Cardiff 1969), pp 247-308.
26 Pennant, T.: *Tours in Wales* (1810), I, p 303.
27 Bingley, W.: *A Tour Round North Wales* (1800), I, p 470.
28 *Report of Commissioners, op. cit.*, pp 140-2.
29 Infra.
30 *The Welshman*, 2 January 1970.
31 An example at the Welsh Folk Museum (Accession Number 69.113) measures 204 inches long, 32 inches deep, with a mesh of 2 inches from knot to knot.
32 *Report of the Commissioners, op. cit.*, p 115.
33 Jones, J. F.: 'Salmon Fisheries 1863', in *The Carmarthenshire Antiquary*, IV (1962-3), p 210.
34 *Ibid.*
35 Donovan, *op. cit.*, II, p 228.
36 I am grateful to Raymond Rees, Secretary of the Tywi Coracle Fishermen's Association, for his assistance in the preparation of this section.
37 *South-West Wales River Board, Fishery Bye-Laws* (Llanelli 1970), p 6.
38 *Ibid.*
39 J. M. Griffiths, Cilgerran, Pembrokeshire. Welsh Folk Museum Record 3309.
40 Raymond Rees of Carmarthen.
41 Hornell, *op. cit.*, p 23.
42 I am grateful to Mr John Thomas, Bonteifi, Cenarth, for a great deal of information on coracle building methods.
43 Hornell, *op. cit.*, p 24.
44 The usual quantity required is 6lb pitch, ½lb linseed oil or with ½lb or less of lard. Oil and lard are added in order to prevent the pitch from becoming brittle and peeling off.
45 Hornell, *op. cit.*, p 26.
46 Wyndham, H. P.: *A Tour through Monmouthshire and Wales . . . 1774 . . . and 1777* [1781], p 86.
47 Malkin, *op. cit.*, II, pp 206-7.
48 The principal casts of Cilgerran fishermen who operated between Tro'r Llyn to within half a mile of Cardigan bridge were – Bwrw byr, Brocen, Gwegrydd, Nantyffil, Crow, Pwlldu, Bwmbwll, Traill Bach. Between Tro'r Llyn and Cwm Sidan, Llechryd coracle men were responsible for fishing. Their principal casts were Bwrw Llyn, Gwddwg, Llyn Ffranc, Pwllglas, Pwlldu, Erllyn, Penllyn, Erglyn,

Dwyfn, Penclaw, Maners, Cor and Dôr.

Between Cwm Sidan and Allt Stradmore, Aber-cuch fishermen fished Ffynnonoer, Maen, Dŵr Bach, Nant-yr-ergyd, Dwfn, Dyfrwyth, Blaen Cyfyn, Pwll Newydd, Fforch, Pig-yr-Agddoe, Dilyn isa, Pwll Dilyn, Dilyn Ucha, and Olchfa.

The Cenarth coracle men operated in Pwll, Gwar, Aberarwen, Dala, Gwar rhyd, Gwar Neudd, Gwar draill, Hen fwlch and Bwlch Bach.

49 Phillips, J. P.: *History of Cilgerran* (1867), p 176.
50 *Ibid.*, p 176.
51 *Ibid.*
52 Oral evidence F. C. Llewelyn. Welsh Folk Museum Record No 111A4.
53 *Minutes of Evidence taken before the Commissioners Appointed to inquire into Salmon Fisheries (England and Wales)* (1861), p 147.
54 A number of Cenarth fishermen refused to pay the licence fee, and some were imprisoned for a month.
55 Phillips, *op. cit.*, p 178.
56 *Minutes of Evidence, op. cit.*, p 141 *et seq.*
57 Salmon Fishery Acts 24 and 25 Vict. c. 109 (1861), 26 and 27 Vict. c. 10 (1863), and 28 and 29 Vict. c. 121 (1865).
58 *Tivy Side Advertizer*, 10 July 1932.
59 Hornell, *op. cit.*, p 13.
60 Bye-Law 6E (Teify and Ayron Fishery Board) 'Prohibitions applicable to particular parts of the Fishery District'.
61 *South-West Wales River Authority Bye-Laws* (1967 ed), p 9.
62 *Report of Committee on Salmon and Freshwater Fisheries* (1964), p 13.
63 *Minutes of Evidence, op. cit.* (1861), p 136.
64 The licence fee for a Tywi coracle in 1972 was £12.00 with endorsement at £0.05 each. On the Teifi coracle licences cost £15.37½ with £0.05 endorsement.
65 Fishery Bye-Laws state 'A coracle net may be used . . . in the River Towy between an imaginary line drawn straight across the said river true north from a signal post adjacent to the main railway 241¼ miles (or thereabouts) from London and an imaginary line drawn straight across the said river from the Railway Pumping Station near the old Carmarthen Tin plate works'.
66 Hornell, *op. cit.*, p 30, states that the following number of coracle licences were issued: 1929 – 25; 1930 – 22; 1931 – 23; 1932 – 19; 1934 – 19; 1935 – 13.
67 *Royal Commission, op. cit.*, p 105.
68 Donovan, E.: *Descriptive Excursions through south Wales and Monmouthshire* (1805), II, p 227.

69 Wyndham, H. P.: *A Tour through Monmouthshire and Wales* (1781), p 52.
70 Evans, J.: *Tour through South Wales* (1804), p 112.
71 *Second Report, op. cit.* (1863), p 22.
72 *Reports on Education in Wales* (HMSO 1847).
73 Williams, D.: *The Rebecca Riots* (Cardiff 1953), p 51.
74 Hornell, *op. cit.*, p 37.
75 *Ibid.*, p 33.
76 *Ibid.*, p 36.
77 *Ibid.*, p 34.
78 Robert Thomas: Welsh Folk Museum Record 3292 Car.
79 *Gwar* the neck of a pool.
80 *Meinen* – a rock on the river bed where nets can be caught (plural – *myneni*).
81 *Gored* – Fish weir.
82 Where the Carmarthen coracle men and Ferryside long netsmen fought in the 1850s.
83 *Pil* – a tidal rivulet.
84 Robert Thomas.
85 *South West Wales River Authority Report* (1970), p 47.
86 *South West Wales River Authority. Proposed Fishery Bye-Laws* (unpublished 1971). Evidence of Bailiff Brian Morgan at Public Enquiry 14 January 1971.
87 *Ibid.* Evidence of Dr W. Roscoe Howells, Water Authority and Fisheries Officer, South West Wales River Authority.
88 *Minutes of Evidence, op. cit.* (1861).
89 Donovan, *op. cit.*, II, p 228.
90 Hornell, *op. cit.*, p 38.
91 R. Rees.
92 Hornell, *op. cit.*, p 40.
93 At the Welsh Folk Museum, St Fagans, the following types of coracle have been preserved. Tywi, Taf, Teifi, Upper Dee, Lower Dee, Conway, Severn (upper), Wye. A Severn coracle (Ironbridge type) has been preserved at the City of Gloucester Museum and a Wye coracle at the Hereford Museum.
94 *Minutes of Evidence, op. cit.*, p 130.
95 Hornell, *op. cit.*, p 41.
96 Miss M. L. Wight of Hereford photographed a number of Cleddau coracles in the nineteen-thirties. Her negatives are now in the collections of the Welsh Folk Museum and provide an irreplaceable record of coracle fishing in west Wales.
97 *Camden's Brittania* translated by E. Gibson. Column 590 [London 1695].

98 Twiston-Davies, L. and Averyl Edwards: *Welsh Life in the Eighteenth Century* (London 1939), p 53
99 Clark, J. H.: *Usk – Past and Present* (Usk no date), pp 161-2.
100 Public Notice Usk and Ebbw Fishery District 1866. Deposited at Brecknock Museum.
101 A coracle at the Hereford City Museum was used on the Wye for fishing by Mr William Dew of Ross-on-Wye, until 1910. It is believed that this was the last coracle in use in the Ross district. They did persist for a little longer in the Monmouth and Redbrook districts.
102 Cox, E.: *Historical Tour through Monmouthshire* (Brecon 1804), p 282.
103 Anonymous: *Travels in Great Britain* (London 1805), II, pp 39-40.
104 *Minutes of Evidence, op. cit.*, pp 30, 33, 44.
105 Gilbert, H. A.: *The Tale of a Wye Fisherman* (London 1929), p 36.
106 Hornell, *op. cit.*, p 264.
107 On loan to the Welsh Folk Museum 1972.
108 Hornell, *op. cit.*, pp 265-6.
109 Hughes-Parry, J.: *A Salmon Fisherman's Notebook* (London, 2nd ed 1955), p 18.
110 *Minutes of Evidence, op. cit.*, p 224.
111 *First Annual Report of the Inspectors of Salmon Fisheries* (London 1862), p 22.
112 *Second Annual Report of the Inspectors of Salmon Fisheries* (London 1863), p 13.
113 *First Annual Report, op. cit.*, p 22.
114 Hughes-Parry, *op. cit.*, p 19.
115 Hornell, *op. cit.*, p 288.
116 *Ibid.*, p 292.
117 *Ibid.*, p 292.
118 *Ibid.*, p 287.
119 See Chapter 5.
120 National Library of Wales MS. 8589 B.
121 Tomos, Dafydd: *Michael Faraday in Wales* (Denbigh 1972), p 95.
122 *Minutes of Evidence, op. cit.*, p 205.
123 Accession number 69.262.
124 Hornell, *op. cit.*, p 272.
125 Coracles were also used on the Worcestershire Avon in the eighteenth century.
126 Hornell, *op. cit.*, p 273.
127 *Ibid.*, pp 274-5.
128 Accession number 37-290/1.
129 Hornell, *op. cit.*, p 277, states 'There are two ways of setting night lines. One is to use a number of lines, each about 15 yards long and

fastened to a stake in the bank. The other is to use a line about 200 yards long and to set it from a coracle zig-zag in the bed of the river'.

130 Davies, A. S.: 'The River Trade and Craft of Montgomeryshire', *Mont. Coll.*, vol 44 (1938-6), p 47.
131 *Ibid.*, pp 178-9.
132 Hornell, *op. cit.*, p 288.
133 Unfortunately no example of a Welshpool coracle has been preserved.
134 Davies, *op. cit. (Mont. Coll.)*, pp 47-8.
135 Donovan, *op. cit.*, II, p 148.
136 Bingley, W.: *A Tour Round North Wales* (1800), Vol I, p 470.
137 *Minutes of Evidence, op. cit.* (1861), p 176.

Chapter 8: Drift Netting (pages 200-22)

1 Davis, F. M.: *An Account of the Fishing Year of England and Wales* (London 1937), p 56.
2 *Ibid.*, p 7.
3 *Report of Committee on Salmon and Freshwater Fisheries* (London 1961), p 145.
4 I am grateful to J. T. Percival, Fisheries Superintendent, Northumbrian River Authority, for a great deal of information on fishing in north-east England.
5 Based on *Northumbrian River Authority Annual Report* (1970), p 42.
6 *Cumberland River Authority Report* (1970), p 28. I am grateful to Mr N. Mackenzie, Fishery Officer, Cumberland River Authority, for assistance.
7 Johnson, R. S.: *River Boards Year Book* (1953), pp 33-4.
8 *East Suffolk and Norfolk River Authority Report* (1971), p 35.
9 Messrs R. Humphreys and J. Kent in 1971.
10 The principal family engaged in drift netting are the Sully family, who moved to Newport from Somerset in the late nineteenth century. I am grateful to Mr C. H. Sully for a great deal of information on the Usk fisheries.
11 I am grateful to Mr Tony Hodsoll of Rhyl for supplying me with information on the Clwyd sling net and its methods of use.
12 *Report of Committee on Salmon and Freshwater Fisheries, op. cit.* (1961), p 145.
13 Messrs Tom Bithell, Peter Bithell, William Bithell and J. E, Bithell. I am grateful to Messrs Tom and Peter Bithell in particular for providing me with information and for allowing me to accompany them on fishing trips.
14 Grimble, *op. cit.*, p 99.
15 A dry extract from the wood of *Acacia catechu.*
16 Davis, F. M., *op. cit.*, p 26.

17 *Wye River Board Fishing Bye-Laws* (1953), p 5. 'Tuck nets shall be nets without bags or pockets and with or without armour. Such nets if unarmoured shall consist of a single sheet or wall of netting as above described having attached round its four edges and on one or both sides a sheet or wall of armour measuring when wet not more than 10 feet in depth and having a mesh of not less than 11 inches from knot to knot or 44 inches round the four sides measured when wet.'
18 In 1971 the licensees were C. J. Sully, W. G. Sully, J. H. Sully, A. Christiansen, J. Chambers, E. J. Holton, C. W. Perkins and I. Wall, all of Newport, Monmouthshire.
19 *Usk River Authority: Fishery Bye-Laws* (1963), p 3.
20 *Ibid.*
21 *Ibid.*
22 Grimble, *op. cit.*, p 193.
23 *Op. cit.*, p 14.
24 *Ibid.*, p 15.
25 Grimble, *op. cit.*, p 194.
26 *Minutes of Evidence . . . Salmon Fisheries, op. cit.*, p 198.
27 I am grateful to Mr Tony Hodsoll of Rhyl for a great deal of information on fishing in the area.
28 *Dee and Clwyd River Board Fishery Bye-Laws* (1966), p 6.
29 Davis, F. M., *op. cit.*, p 64.
30 *Second Report of Commissioners, op. cit.*, p 20.
31 *Ibid.*
32 *Minutes of Evidence, op. cit.*, p 220.
33 *Second Report, op. cit.*, p 20.
34 These figures refer to 1866. In 1871, 57 draft net and 27 coracle net licences were issued. By 1884 the number of draft net licences, at £5 each, had increased to 85 and coracle licences at £2.5s each had declined to 15.
35 Grimble, *op. cit.*, p 202.
36 *Ibid.*, p 204.
37 Davis, F. M., *op. cit.*, p 53.
38 *Ibid.*, p 53.
39 *Dee and Clwyd River Board Fishery Bye-Laws* (1961 ed), p 6.
40 Davis, F. M., *op. cit.*, p 7.
41 Messrs Taylor of Chester.
42 Licence T.1. P. Bithell with C. Bithell; P. Bithell Jun; T. Stealey.
Licence T.2. Wm. Bithell with J. Bithell; C. Thompson.
Licence T.3. J. E. Bithell with T. Johnson; G. Williams.
Licence T.4. T. Bithell with E. Bithell and R. Bithell.

Chapter 9: Seine Netting (pages 223-62)

1 Also known as 'a draft net', 'draw net' and 'long net'.
2 Davis, F. M., *op. cit.*, p 66.
3 I am grateful to the officers of the various river authorities for these statistics.
4 Mesh is measured from knot to knot, the measurement being taken when the net is wet.
5 The figures are based on Fishery Regulations issued by the South West Wales, Gwynedd and Dee and Clwyd River Authorities.
6 Taylor, J. N.: *Illustrated Guide to the Severn Fishery Collections* (Gloucester 1953), p 4.
7 Owen, G.: *The Description of Pembrokeshire* (1892 ed), II, p 117. In a footnote Henry Owen, the editor of the 1892 edition, says of the word *seine* 'Compare with the French *seine* or *senne*, anciently *seîne* (from Latin *sagena*) – a drag net.'
8 *First Annual Report, op. cit.*, p 27.
9 Grimble, *op. cit.*, p 153.
10 *Op. cit.*, p 20.
11 I am grateful to Messrs James Sallis and Washington Thomas in particular for a great deal of information on seine netting in the Teifi.
12 The shots are usually flat pebbles with numbers 1 to 5 written on in chalk, although in 1971 it was noted that bingo discs had replaced the traditional stones. An early 19th-century document (N.L.W.5603B) describes the Netpool as 'the plot of greensward where fishermen dry their nets; the battleground of fish women; the arena of the war of tongues'.
13 A *gwrhyd*, ie *gwr* (man) + *hyd* (length), signifies the length that can be measured by outstretching a man's arms. A *gwrhyd* is approximately 2 yards.
14 Mr James Sallis: Welsh Folk Museum, Department of Oral Tradition and Dialect Record No 3300 (1971).
15 *South West Wales River Authority, Fishery Bye-Laws* (1968). In 1970 the six licensees who pay an annual fee of £28.00 caught 802 salmon (average weight of 9½lb) and 254 sewin (average weight 3¼lb).
16 James Sallis. Record 3301 (Welsh Folk Museum).
17 Roberts, Gomer: 'Myfyrdod ar Garreg y Fendith', *Cardigan & Tivy Side Advertiser*, 9 March 1962.
18 Anson, Peter F.: *Fisher Folklore* (Faith Press 1965), pp 103-4. '. . . all seafarers have believed it unlucky to meet a clergyman of any Christian denomination when on their way to their vessels, or to see one standing near them. This is an ancient and almost universal superstition . . . It does not appear to be associated with any anti-clerical feeling'.

19 There has been a considerable decline in catches in the Teifi estuary in recent years. In 1969, for example, the Teifi seine nets caught 507 salmon and 186 sewin while coracle nets further upstream caught 110 salmon and 309 sewin, a total of 617 salmon and 495 sewin. This compares with 1,185 salmon and 685 sewin caught in 1959, and 1,569 salmon and 817 sewin in 1940. During the first decade of the present century, when recording of catches began, a total of between 3,000 and 5,000 salmon caught by nets was commonplace.
20 *South West Wales River Board, Fishery Bye-Laws* (1967), p 10.
21 *Minutes of Evidence taken before the Commissioners appointed to Inquire into Salmon Fisheries (England and Wales)* (1861), p 117.
22 *Ibid.*, p 117.
23 Welsh Folk Museum Record No 4291 (Robert Thomas, Carmarthen).
24 *Second Report, op. cit.*, p 22.
25 *Minutes of Evidence, op. cit.*, p 110.
26 *Ibid.*, p 118.
27 I am grateful to Mr Elwyn Brown of Laugharne, the sole licensee, for his assistance and in supplying me with information on the methods of wade netting.
28 *Second Annual Report, op. cit.*, p 22.
29 *South-West Wales River Board, Fishery Bye-Laws.*
30 *South-West Wales River Board, Fishery Bye-Laws* (1967), p 5.
31 *Minutes of Evidence taken before the Commissioners appointed to Inquire into Salmon Fisheries* (1861), p 134.
32 *Second Annual Report of the Inspectors of Salmon Fisheries (England and Wales)* (1863), p 20.
33 Frank Thomas, Newport. Welsh Folk Museum Record 3298.
34 *Second Annual Report, op. cit.*, p 21.
35 *Ibid.*, p 21.
36 Grimble, *op. cit.*, p 148.
37 Grimble, *op. cit.*, p 163.
38 *Second Annual Report, op. cit.*, p 18.
39 Grimble, *op. cit.*, p 163.
40 *Op. cit.*, p 17.
41 *Ibid.*, p 17.
42 *Minutes of Evidence, op. cit.*, p 184.
43 Welsh Folk Museum Record No 3295 (William Jones, Porthmadog).
44 Licensees in 1969 were J. Roberts; W. H. Jones; W. O. Jones; W. Hughes; J. D. Craven and Llewelyn Jones, all of Conway.
45 *Minutes of Evidence, op. cit.*
46 *Minutes of Evidence, op. cit.*, p 226.
47 Waters, Brian: *Severn Tide* (London 1947), p 54.

48 *Ibid.*, p 38.
49 Johnson, R. S.: 'Morecambe Bay Salmon' in *The River Boards' Association Year Book*, No 1 [1953], p 32.
50 *Ibid.*

Chapter 10: Eel Capture (pages 263-87)

1 Taylor, J. N.: 'Elver Fishing on the River Severn' in *Folk Life*, Vol III (1965), p 55.
2 *Ibid.*
3 On the Severn silver eels are called 'right eels' or 'vawsen'.
4 *Capture of Eels* (Ministry of Agriculture & Fisheries, 1954), pp 1-2.
5 I am grateful to J. Neufville Taylor, Curator of the City of Gloucester Museums, for allowing me to quote freely from his work on elver fishing in the Severn estuary.
6 Severn River Authority, Sixth Annual Report (1971), p 29.
7 Taylor, *op. cit.* (1965), p 56.
8 *Ibid.*
9 39 & 40 Victoria, c. 34.
10 25 Henry VIII, c. 7.
11 18 George III, c. 33.
12 36 & 37 Victoria, c. 71.
13 Waters, I.: *About Chepstow* (Newport 1952), p 39.
14 Camden: *Brittania* (1722 ed), Vol I, p 96.
15 Waters, *op. cit.*, p 7.
16 I am grateful to Mr G. H. Bielby, Fisheries Officer, Cornwall River Authority, for information on Gunnislake Weir.
17 Porter, E. M.: 'The Cambridgeshire Fens' in *Gwerin*, Vol II, No 1, p 15, says 'Fenmen and women believed that the wearing of eelskin garters protected them from rheumatism and ague. The skins were hung in the sun until they were dry and stiff, when they were tied at the ends, greased with fat and worked over a round piece of wood until they were pliable again. The ends were then untied and the skins stuffed with chopped thyme and lavender leaves and placed in a linen bag between layers of freshly gathered marsh mint. The bag was buried for most of the summer under the peat. When the skins were taken out, the thyme and lavender were removed and a further polish given to the skins with a smooth stone.'
18 Cf. *sticher* – fish spear.
19 *stang* – a variant of *sting* (Wright, J.: *The English Dialect Dictionary* (1898-1905), Vol V, p 727. Nevertheless the word *stang* may be a variant of *stong* (Old Norse) – pole.
20 *gad* – *goad* (c. pole).

21 *pil* – pointed stick + *gar* – spear.
22 Green, Charles: 'Eel Spears' in *Antiquity*, Vol XXII [1948], pp 14-18.
23 *Ibid*., p 14.
24 *Ibid*., p 16.
25 Taylor, *op. cit.*, p 10.
26 Dodgson, J. McN.: 'H. B. Dodgson on Fishing at Preesall, Lancashire', unpublished MS [1955].
27 *Eel Capture, Fisheries Notice* (HMSO 1954), p 14.
28 Taylor, *op. cit.*, p 9.
29 Dodgson, *op. cit.*, p 14.
30 Peate, I. C.: 'Severn Eel Traps', in *Man*, Vol XXXIV [1934], p 178.
31 Waters, B.: *Severn Tide* [London 1947], p 43.
32 Tebbutt, C. F. and Sayce, R. U.: 'Fenland Eel Traps', in *Man*, Vol XXXVI [1936], p 179.
33 Sayce, R. U.: 'Traps and Snares', in *The Montgomeryshire Collections*, Vol XLIX [1945], Part 1, p 72.
34 Peate, *op. cit.*, p 178.
35 Taylor, *op. cit.*, p 9.
36 *Ibid.*, p 19.
37 *Ibid.*, p 19.
38 Bristol Museum MSS. Royal 10E IV.
39 Peniarth 56, National Library of Wales.
40 *Report of the Commissioners Appointed to Inquire into Salmon Fisheries (England and Wales)*, p xxiv.
41 Davis, *op. cit.*, p 41.
42 *East Suffolk and Norfolk River Authority Sixth Annual Report* [1971], p 41, gives the following details of licences issued: 'Trap not exceeding 4ft – 79; Spear – 2; Dutch fyke net – 457; Trap at mill – 5; Line with more than one hook – 36; Drag net – 6; Beam trawl – 2; Dutch or Belgian square net – 1.'
43 *Capture of Eels, op. cit.*, p 3.
44 Burgess, John: *Eel Trapping* (Bridport 1971), p 6.
45 *Capture of Eels, op. cit.*, p 9.
46 Davis, *op. cit.*, p 42.
47 Waters, *op. cit.*, p 42.
48 Davis, *op. cit.*, p 41.
49 Burgess, *op. cit.*, p 8.

Chapter 11: Poaching (pages 288-304)

1 I am particularly grateful to Mr R. I. Millichamp, E.R.D. Fishing Officer to Usk River Authority, for his assistance in compiling this chapter.

2 *Minutes of Evidence, op. cit.*, p 105.
3 *Ibid.*, p 224.
4 *Ibid.*, pp xv-xvi.
5 *Sunday Times*, 8 December 1968.
6 *Minutes of Evidence, op. cit.*, p 33.
7 Williams, James: *Brecon and its Neighbourhood* (n.d.), pp 18-20.
8 Fenton, A.: 'Scottish Salmon Fishing Spears', *The Salmon Net*, No 4 (June 1968), pp 31-46.
9 The Welsh for a salmon spear is *tryfer*, other synonyms are *leister; waster; vanch spear; dart; prick; fasher.*
10 25 Henry VIII C. 7.
11 In the Welsh Folk Museum collection there are over a 100 salmon spears, most of them made in the nineteen-fifties and sixties, and all confiscated by the river authorities in recent years.
12 Fenton, *op. cit.*, p 34.
13 Williams, D. J.: *The Old Farmhouse* (London 1961), p 150.
14 Welsh Folk Museum MSS 1101.
15 Sayce, R. U.: 'The Otter in Wales and north-west Europe', *Mont. Coll.*, Vol 53 (1953-4), pp 140-54.
16 *Ibid.*, p 147.

Select Bibliography

ANSON, P. F., *Fisher Folklore,* London (1965).

ANSON, P. F., *Scots Fisherfolk,* Saltaire Society (1950).

ASSOCIATION OF RIVER AUTHORITIES – *ANNUAL YEAR BOOKS.*

CHALLENDER SOCIETY, *The Science of the Sea,* London (1912:1927).

CUNNINGHAM, J. T., *The Preservation of Fishing Nets,* London (1902).

DAVIES, D. C., 'The Fisheries of Wales', *Transactions Liverpool National Eisteddfod 1884,* Liverpool (1885), pp 285-320.

DAVIES, H. R., *The Conway and the Menai Ferries,* Cardiff (1942).

DAVIS, F. M., *An Account of the Fishing Gear of England and Wales,s,* London (4th ed 1958).

EVERY, S. F., *The Art of Netting,* London (1845).

FRASER, R., *A Review of the Domestic Fisheries of Great Britain,* Edinburgh (1818).

GIBBS, W. E., *The Fishing Industry,* London (1922).

GRIMBLE, S. A., *The Salmon Rivers of England and Wales,* London (c 1905).

HMSO, *Report of Committee on Salmon and Freshwater Fisheries,* London (1961).

HMSO, *Report of the Commissioners appointed to Inquire into Salmon Fisheries (England and Wales),* London (1861).

HMSO, *Reports of the Inspectors of Salmon Fisheries (England and Wales),* London (1862, 1863).

HOLDSWORTH, E. W. H., *Apparatus for Fishing,* London (1883).

HOLDSWORTH, E. W. H., *Deep Sea Fishing Boats,* London (1874).

HORNELL, J. A. S., *British Coracles and Irish Curraghs,* London (1938).

IEBC, *Fishing and Shooting,* London (1874).

JENKINS, J. G., 'Commercial Salmon Fishing in Welsh Rivers', *Folk Life,* Vol 9 (1971), pp 29-60.

JENKINS, J. G., 'The Customs of Welsh Fishermen', *Folklore,* Vol 83 (1972), pp 1-19.

JENKINS, J. T., *The Fishes of the British Isles both Fresh Water and Salt*, London (1925).

JENKINS, J. T., *The Sea Fisheries*, London (1920).

JOHN, D., *Flyfishing on the Usk*, Brecknock Museum (1968).

JOHNSTONE, J., *British Fisheries*, London (1905).

JONES, A. M., *The Rural Industries of England and Wales*, Vol IV, Oxford (1927).

KITE, OLIVER, *A Fisherman's Diary*, London (1969).

LAMBETH, R. C., *Some Former Cambridgeshire Agricultural and Other Implements*, Reprint from *East Anglian Magazine*, Cambridge (1940).

LEWIS, E. A., *The Welsh Port Books 1550-1603*, London (1927).

LOWE, W. BEZANT, *The Heart of Northern Wales*, Llanfairfechan (1929).

PATERSON, A. H., *Man and Nature*, London (1858).

PHILLIPS, J. R., *The History of Cilgerran*, London (1867).

POLLOCK, F., 'The Fishery Laws', *International Fisheries Exhibition Literature*, Vol 1, London (1884).

READY, OLIVER, *Life and Sport on the Norfolk Broads*, London (c 1911).

SALZMAN, L. F., *English Industries and the Middle Ages*, Oxford (1923).

SENOGLES, D., *The Story of Ynys Gorad Goch*, Menai Bridge (1969).

SHEPPARD, T., *Catalogue of the Museum of Fisheries and Shipping*, Hull (1938).

TAYLOR, J. N., *Guide to the Severn Fishery Collection*, Gloucester (1949).

THOMSON, D., *The Seine Net*, London (1969).

TUCKER, N., *Conway and its Story*, Denbigh (1960).

WALPOLE, S., 'The British Fish Trade', *International Fisheries Exhibition Literature*, Vol 1, London (1884).

WATERS, BRIAN, *Severn Tide*, London (1947).

WHITE, P. J., *The Sea Fisheries of North Wales*, Liverpool (nd).

Index